ISBN 978-1-330-36958-6
PIBN 10041968

English
Français
Deutsche
Italiano
Español
Português

www.forgottenbooks.com

Mythology Photography **Fiction**
Fishing Christianity **Art** Cooking
Essays Buddhism Freemasonry
Medicine **Biology** Music **Ancient**
Egypt Evolution Carpentry Physics
Dance Geology **Mathematics** Fitness
Shakespeare **Folklore** Yoga Marketing
Confidence Immortality Biographies
Poetry **Psychology** Witchcraft
Electronics Chemistry History **Law**
Accounting **Philosophy** Anthropology
Alchemy Drama Quantum Mechanics
Atheism Sexual Health **Ancient History**
Entrepreneurship Languages Sport
Paleontology Needlework Islam
Metaphysics Investment Archaeology
Parenting Statistics Criminology
Motivational

APPLETONS' HOME READING BOOKS

THE

STORY OF THE AMPHIBIANS

AND

THE REPTILES

BY

JAMES NEWTON BASKETT

AUTHOR OF THE STORY OF THE FISHES
THE STORY OF THE BIRDS, ETC.

AND

RAYMOND L. DITMARS

CURATOR OF REPTILES AT THE NEW YORK ZOOLOGICAL PARK

NEW YORK
D. APPLETON AND COMPANY
1902

Published June, 1902

TO

MY SON

HOWARD GORDON BASKETT

A LOVER OF

THE HUMBLER CREATURES

INTRODUCTION TO THE HOME READING BOOK SERIES BY THE EDITOR

THE new education takes two important directions—one of these is toward original observation, requiring the pupil to test and verify what is taught him at school by his own experiments. The information that he learns from books or hears from his teacher's lips must be assimilated by incorporating it with his own experience.

The other direction pointed out by the new education is systematic home reading. It forms a part of school extension of all kinds. The so-called "University Extension" that originated at Cambridge and Oxford has as its chief feature the aid of home reading by lectures and round-table discussions, led or conducted by experts who also lay out the course of reading. The Chautauquan movement in this country prescribes a series of excellent books and furnishes for a goodly number of its readers annual courses of lectures. The teachers' reading circles that exist in many States prescribe the books to be read, and publish some analysis, commentary, or catechism to aid the members.

Home reading, it seems, furnishes the essential basis of this great movement to extend education beyond the school and to make self-culture a habit of life.

Looking more carefully at the difference between the two directions of the new education we can see what each accomplishes. There is first an effort to train the original powers of the individual and make him self-active, quick at observation, and free in his thinking. Next, the new education endeavors, by the reading of books and the study of the wisdom of the race, to make the child or youth a participator in the results of experience of all mankind.

These two movements may be made antagonistic by poor teaching. The book knowledge, containing as it does the precious lesson of human experience, may be so taught as to bring with it only dead rules of conduct, only dead scraps of information, and no stimulant to original thinking. Its contents may be memorized without being understood. On the other hand, the self-activity of the child may be stimulated at the expense of his social well-being—his originality may be cultivated at the expense of his rationality. If he is taught persistently to have his own way, to trust only his own senses, to cling to his own opinions heedless of the experience of his fellows, he is preparing for an unsuccessful, misanthropic career, and is likely enough to end his life in a madhouse.

It is admitted that a too exclusive study of the knowledge found in books, the knowledge which is aggregated from the experience and thought of other people, may result in loading the mind of the pupil with material which he can not use to advantage.

Some minds are so full of lumber that there is no space left to set up a workshop. The necessity of uniting both of these directions of intellectual activity in the schools is therefore obvious, but we must not, in this place, fall into the error of supposing that it is the oral instruction in school and the personal influence of the teacher alone that excites the pupil to activity. Book instruction is not always dry and theoretical. The very persons who declaim against the book, and praise in such strong terms the self-activity of the pupil and original research, are mostly persons who have received their practical impulse from reading the writings of educational reformers. Very few persons have received an impulse from personal contact with inspiring teachers compared with the number that have been aroused by reading such books as Herbert Spencer's Treatise on Education, Rousseau's Émile, Pestalozzi's Leonard and Gertrude, Francis W. Parker's Talks about Teaching, G. Stanley Hall's Pedagogical Seminary. Think in this connection, too, of the impulse to observation in natural science produced by such books as those of Hugh Miller, Faraday, Tyndall, Huxley, Agassiz, and Darwin.

The new scientific book is different from the old. The old style book of science gave dead results where the new one gives not only the results, but a minute account of the method employed in reaching those results. An insight into the method employed in discovery trains the reader into a naturalist, an historian, a sociologist. The books of the writers above named have done more to stimulate original research on the part of their readers than all other influences combined.

It is therefore much more a matter of importance to get the right kind of book than to get a living teacher. The book which teaches results, and at the same time gives in an intelligible manner the steps of discovery and the methods employed, is a book which will stimulate the student to repeat the experiments described and get beyond them into fields of original research himself. Every one remembers the published lectures of Faraday on chemistry, which exercised a wide influence in changing the style of books on natural science, causing them to deal with method more than results, and thus train the reader's power of conducting original research. Robinson Crusoe for nearly two hundred years has aroused the spirit of adventure and prompted young men to resort to the border lands of civilization. A library of home reading should contain books that incite to self-activity and arouse the spirit of inquiry. The books should treat of methods of discovery and evolution. All nature is unified by the discovery of the law of evolution. Each and every being in the world is now explained by the process of development to which it belongs. Every fact now throws light on all the others by illustrating the process of growth in which each has its end and aim.

The Home Reading Books are to be classed as follows:

First Division. Natural history, including popular scientific treatises on plants and animals, and also descriptions of geographical localities. The branch of study in the district school course which corresponds to this is geography. Travels and sojourns in distant lands; special writings which treat of this or that animal or plant, or family of animals or plants; anything that relates to organic nature or to meteorology, or descriptive astronomy may be placed in this class.

Second Division. Whatever relates to physics or natural philosophy, to the statics or dynamics of air or water or light or electricity, or to the properties of matter; whatever relates to chemistry, either organic or inorganic—books on these subjects belong to the class that relates to what is inorganic. Even the so-called organic chemistry relates to the analysis of organic bodies into their inorganic compounds.

Third Division. History, biography, and ethnology. Books relating to the lives of individuals; to the social life of the nation; to the collisions of nations in war, as well as to the aid that one nation gives to another through commerce in times of peace; books on ethnology relating to the modes of life of savage or civilized peoples; on primitive manners and customs—books on these subjects belong to the third class, relating particularly to the human will, not merely the individual will but the social will, the will of the tribe or nation; and to this third class belong also books on ethics and morals, and on forms of government and laws, and what is included under the term civics, or the duties of citizenship.

Fourth Division. The fourth class of books includes more especially literature and works that make known the beautiful in such departments as sculpture, painting, architecture and music. Literature and art show human nature in the form of feelings, emotions, and aspirations, and they show how these feelings lead over to deeds and to clear thoughts. This department of books is perhaps more important than any other in our home reading, inasmuch as it teaches a knowledge of human nature and enables us to understand the motives that lead our fellow-men to action.

PLAN FOR USE AS SUPPLEMENTARY READING.

The first work of the child in the school is to learn to recognize in a printed form the words that are familiar to him by ear. These words constitute what is called the colloquial vocabulary. They are words that he has come to know from having heard them used by the members of his family and by his playmates. He uses these words himself with considerable skill, but what he knows by ear he does not yet know by sight. It will require many weeks, many months even, of constant effort at reading the printed page to bring him to the point where the sight of the written word brings up as much to his mind as the sound of the spoken word. But patience and practice will by and by make the printed word far more suggestive than the spoken word, as every scholar may testify.

In order to bring about this familiarity with the

printed word it has been found necessary to re-enforce the reading in the school by supplementary reading at home. Books of the same grade of difficulty with the reader used in school are to be provided for the pupil. They must be so interesting to him that he will read them at home, using his time before and after school, and even his holidays, for this purpose.

But this matter of familiarizing the child with the printed word is only one half of the object aimed at by the supplementary home reading. He should read that which interests him. He should read that which will increase his power in making deeper studies, and what he reads should tend to correct his habits of observation. Step by step he should be initiated into the scientific method. Too many elementary books fail to teach the scientific method because they point out in an unsystematic way only those features of the object which the untutored senses of the pupil would discover at first glance. It is not useful to tell the child to observe a piece of chalk and see that it is white, more or less friable, and that it makes a mark on a fence or a wall. Scientific observation goes immediately behind the facts which lie obvious to a superficial investigation. Above all, it directs attention to such features of the object as relate it to its environment. It directs attention to the features that have a causal influence in making the object what it is and in extending its effects to other objects. Science discovers the reciprocal action of objects one upon another.

After the child has learned how to observe what is essential in one class of objects he is in a measure fitted to observe for himself all objects that resemble this class.   After he has learned how to observe the seeds of the milkweed, he is partially prepared to observe the seeds of the dandelion, the burdock, and the thistle.   After he has learned how to study the history of his native country, he has acquired some ability to study the history of England and Scotland or France or Germany.   In the same way the daily preparation of his reading lesson at school aids him to read a story of Dickens or Walter Scott.

The teacher of a school will know how to obtain a small sum to invest in supplementary reading.   In a graded school of four hundred pupils ten books of each number are sufficient, one set of ten books to be loaned the first week to the best pupils in one of the rooms, the next week to the ten pupils next in ability. On Monday afternoon a discussion should be held over the topics of interest to the pupils who have read the book.   The pupils who have not yet read the book will become interested, and await anxiously their turn for the loan of the desired volume.   Another set of ten books of a higher grade may be used in the same way in a room containing more advanced pupils. The older pupils who have left school, and also the parents, should avail themselves of the opportunity to read the books brought home from school.   Thus is begun that continuous education by means of the public library which is not limited to the school period, but lasts through life.   W. T. Harris.

Washington, D. C., *Nov. 16, 1896.*

PREFACE.

THE average reader, old or young, does not usually find himself so much interested in an amphibian or reptile as he does in fishes, birds, or mammals, because they are not often objects of pursuit for either "sport" or food. In fact, casually, they are abhorrent. But if he should be one of those whose interest goes beyond that of the mere amusement which satisfies the most primitive of his instincts, he will nowhere in the realm of animal life find objects more worthy of his attention. Herein Nature, with the potter's clay of plastic things in her palms, seemed to have tarried in delightful experiment before she shaped the higher and better creatures; and in the amphibians especially —even more so than the fishes—appears to have indulged every passing caprice and suggestion.

To look in on her in some of her vagaries, and note her as she seems to put, drop by drop, the alchemy of change into the fuming elements, is partly the object of this little volume. The author indulges the hope, also, that the humble, creeping things herein described may not be longer despised, but that a more intimate knowledge of them will help to arouse a sympathetic interest in one of the ostracised families of the animate world.

J. N. B.

CONTENTS

PART I.—AMPHIBIANS

PART II.—REPTILES

PART III.—A COLLECTOR'S EXPERIENCE WITH REPTILES

LIST OF ILLUSTRATIONS

xxi

PART I

AMPHIBIANS

By JAMES NEWTON BASKETT

STORY OF THE AMPHIBIANS

CHAPTER I

WHAT AMPHIBIAN MEANS

THE term Amphibians is used to designate that great class of the backboned animals, which includes the Frogs, Toads, Salamanders, Mudpuppies, etc. Unfortunately there is no good English word for all these, as there is for the fishes and for the birds, or no good Anglicized word as those for the reptiles and for the mammals.

We are a little apt to confuse amphibian with the amphibious; but the latter term is used loosely to define any creature capable of staying for indefinite periods either in water or air—such as may have two abiding-places. But an amphibian is a creature having, usually and normally, as it grows, two *forms* of life. Thus seals, otters, muskrats, and beavers are often spoken of as amphibious, but they are really mammals; while nearly all true amphibians, such as frogs, salamanders, etc., have a tadpole state through which they pass in their growth, and in which they are entirely water-haunting; and later they have an adult stage in which they may be either land-haunters purely, water-haunters, or amphibious, like those mammals mentioned. Having two lives is the original meaning of amphibious (from the Greek, *amphi*, two, and *bios*, life).

Some writers speak of this class of the vertebrates as the Batrachians (from Greek *batrachios*, a frog), but the author prefers to leave this term as the scientific name of the tailless amphibians only.

There are a few fishes that, when young, have a tadpole state, but these, when grown, are easily distinguished from any amphibian, either because they have no true legs, or because they have very distinct fins. But there are many tadpoles of the amphibians, which outwardly resemble certain fishes, and close examination is required and technical terms must be used to distinguish them.

In a general way, every tadpole is a low order of fish, having gills and living a thoroughly aquatic life, but later they all either acquire true limbs with toes, or else they have better lungs than has any fish. There are fishes with lungs, but no legs. At the same time there are many amphibians with legs, that still retain the gills of their tadpole state and have a very poor sort of lungs indeed.

From the classes above them the tadpole condition of the amphibians is the most characteristic distinction, though some do not have this larval condition outside of the egg. This egg is also quite different from that of reptiles and birds, and in the process of hatching, the tadpole is not enclosed in certain sacs or membranes, which grow round the young of the other two classes and nourish them.

Anatomically, the rule is that amphibians have no such complete ribs as are found in the other classes. Externally, it may be said in a general way of those living now, that amphibians are naked-skinned, reptiles are scaly, mammals have hair and birds have feathers. Some reptiles, however, are not, and a few mammals are scaly.

FIG. 1.—Frog (*Discoglossus pictus*) in action, showing free development of limbs as compared with other members of its class.

Likewise many extinct amphibians and reptiles had paddle-shaped limbs. Their structure, however, shows that these were not true fins, but were made out of a true three-jointed leg. Likewise some of each of these classes have no legs at all, but they are readily distinguished by the other characters noted.

Amphibians were once classed with reptiles, because both were cold-blooded creeping things; but although a lizard and a salamander may look much alike, mere resemblance is no longer a basis of classification. Kinship is a matter of structure.

While the fishes were the first creatures to have a backbone, the amphibians were the first to take this great weapon and go out to conquer the dry land.

Fig. 2.—Giant salamander (*Megalobatrachus maximus*).

They were thus the pioneers of all the reptiles and mammals, which have since subdued the earth, and of the birds which have invaded the air. With them came in the three-jointed limb and the fingers and the toes. The many fringes of the fins of the fishes

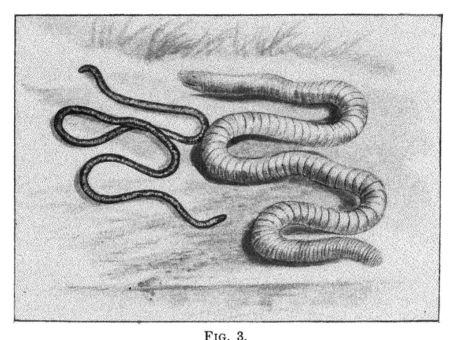

FIG. 3.

Slender cæcilia   White-bellied cæcilia
(*Cæcilia gracilis*).   (*Cæcilia lentaculata*).

rapidly decreased to five digits in this next class, and then the human hand lay in its cradle among the rushes—a giant which should rise and strike and strangle.

With the jointed limb and toe came in the lung, also, in its best development—permitting the excursion away from the water.

There are now living three orders of amphibians, easily distinguished from each other. First there are the *tailed* forms, like the salamanders, always having a tail and at least two limbs. Second there are the *tailless* forms, such as frogs and toads, always having four limbs. Third there are the legless forms, the Cæcilians, which have no perfect limbs at all, though stumps show in the very young. They are wormlike in shape—are burrowing creatures and are practically eyeless.

In the long-ago there lived many other kinds of amphibians.

CHAPTER II

LIMBS AND TOES

THE typical form of each order is illustrated in
the last chapter. In place of a tail the frogs have
their hind limbs capable of stretching out directly in
line with the body. This gives them a great thrust
in leaping and swimming, and the long legs thus trailing act like a feather on an arrow in one case, and
like a rudder in the other. The fore legs of the tailless forms are weak, and are used mostly in alighting
and in propping up the forepart of the body. These
nearly all leap.

In the tailed forms, the legs are all usually rather
weak, and there is no great difference in the size of
the fore and hind pairs, as there is in the frogs. To
this order the forward pair seems the more important
since they serve to drag the creature slowly along, and
they are never lost, though the rear ones are gone in
the sirens. So, also, the fore limbs first develop in
their tadpoles, while in those of the frogs the rear
limbs show first. In one tailed form, known as the
Congo snake (though it is not a snake), all four of the

3   7

limbs are small and useless.   The creature moves by wriggling.   (See Fig. 4.)

FIG. 4.—Congo snake (*Murœnopsis tridactyla*).

The Cæcilians have bands around the body, by which they pull themselves through the ground.

In the legged forms the number of the toes varies.

In the grasping hand of the frogs and tree-toads, there is found the first thumb in nature.   So perfect is this that many tree-toads can suspend themselves for some time by a single hand.   These have soft round pads on the ends of the digits, which enable them to stick to smooth surfaces—the slipping being prevented by moisture.   The cricket-frog can, by the mere adhesion of its moist underparts, climb up vertical glass and remain there even when so turned that its back is downward; and one little salamander, having neither pads nor claws, can so run on ceilings.

Claws are very poorly developed in the amphibians. Some of the tailed forms have horny tips on their toes all the time; others have these at certain seasons only, when they chase or grasp each other. These latter are on the fore feet, and are shed later. In Africa, there is a frog armed with spurlike claws on three toes of the hind feet, and our own spade-foot toad (Fig. 5) has a flat spur on its rear foot, which is evidently used as a burrowing implement.

The feet of the so-called Surinam toad (Fig. 6) (*Pipa*) are tipped with a starlike sprangle. The rule

FIG. 5.—Spadefoot frog (*Scaphiopus holbrookii*).

among frogs and toads is that the rear toes are webbed and the front ones are not. The length of rear toes and the extent of the webs vary much.

It is said that in one tree-toad of Borneo, the usual disks are so large, and the membranes between them so broad, that when the toes are spread, the creature

FIG. 6.—Surinam toad (*Pipa Americana*), tadpole on the left.

may sail from tree to tree on them, after the manner of the flying squirrels.   (See Fig. 7.)

In the European newt, a dry-land tailed form, the males develop webs at that season only when all go to the water; and these dry up and fall off when they go back to the land.

Tails

Of course when in the water the tailed forms swim largely by means of their tails, and in those which remain there most of the time the tail is flat vertically, like that of a fish, and has a fringe on the upper edge, like a fin. But in those which are almost purely land-haunting, as the true salamanders, the tail is round. But some amphibians, also, have a fringe on the tail at that time only when they come to the water. In none of these fins now are there any supporting filaments or spines, as there are in those of the fishes; and no amphibian has fins on its sides.

Tongues

In this class of vertebrates the tongue is an interesting member, and here finds its first and best

FIG. 7.—Flying tree-frog (*Rhacophorus reinhardtii*).

development (Fig. 8). In some frogs it is entirely absent, but in most of them it is large, and can be thrust out very far as a capturing instrument—having usually a sticky substance on its end. In the tailed amphibians, the tongue is variable, and quite helpful in describing groups. In some it is a mere wrinkling of a membrane on the floor of the mouth. It is so in the giant salamanders. In no case is it ever split, nor is it thread-like and capable of being thrust far out while the mouth is closed, as it is in some reptiles. Legless amphibians may always be thus known from legless reptiles.

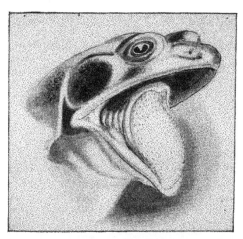

FIG. 8. — Head of frog, showing tongue fixed in front but free behind.

In the tailed amphibians, the tongue is not free behind and tied in front only, as in frogs, but in a few salamanders it is free all around and tied in the center. This freedom may be so great that there is left only a central stem (or pedestal), and the whole becomes mushroom-shaped. In a few cases this pedestal is capable of stretching, so that the cap may perhaps be thrust out of the mouth. But since above this kind of tongues there is usually a quantity of teeth on the roof of the mouth, it is not unlikely that the tongue is used to grind the food against these. In other salamanders, the tongue is free at the sides, but only so in a limited degree behind. In the sirens, which have no teeth, it is free in front to a slight extent. A peculiarity of one American genus (*Ambly-stoma*) is that the tongue is pleated or wrinkled on top, and the folds or creases run from some point within outward, as the spokes of a wheel. This point may be in the center or toward the rear; and its position aids in distinguishing species. Here as elsewhere the tongue is helpful in " diagnosing the case."

In the Cæcilians the tongue is like that of the salamander forms—fixed to the floor of the mouth—and can not be thrust out, as in legless reptiles.

CHAPTER III

TEETH

WITH the amphibians of to-day teeth seem to be
of less importance than in any other class of backboned creatures. Even fishes have developed them
much more terribly. But there were once fierce
amphibians which had great teeth; and because
these, when cut across their length, showed markings made by the folds, which resembled labyrinths,
these old monsters are called Labyrinthodonts (Fig.
9). The horned toad of Brazil, still has these in-
folded teeth, with grooves on the sides. In reptiles
having such grooved teeth there is always found a
poison ready to flow down them, and so it is said that
this wrinkled-toothed toad bites viciously, pursues its
enemies, and has poisonous teeth.

While the amphibians may have teeth elsewhere
than on the jaws, none have them on the tongue,
as do many fishes. The labyrinthodonts had great
tusks in the throat, but in many modern forms teeth
may be absent from either or both jaws. In the
United States common toads have no teeth on the

14

jaws, but this is not true of toads everywhere. The tree-toads (which are toads not frogs) have teeth on the upper jaw, and some real tree-frogs (*Dendrobatidæ*) have no teeth on either jaw. In the tailed amphibians there are usually teeth on both jaws, but the siren has a beak only, like that of turtles and some fishes. Tadpoles of the frogs and toads have similar beaks. In the Cæcilians there are teeth on both jaws— especially the lower.

FOOD AND FEEDING HABITS

In the grown-up state, when not confined, all amphibians appear to be either flesh-eaters or insect-eaters. In confinement many tailed forms will eat bread and milk, or bread alone, and other cooked forms of vegetable food. But their

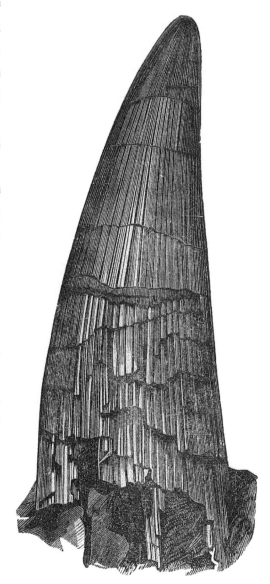

FIG. 9.—Tooth of labyrinthodont, natural size.

tadpoles are, to a large extent, vegetable eaters—living on grass and reeds. The tadpoles of the spadefoot toad are said to be especially fond of each other, and many are otherwise carnivorous. Such creatures as ducklings, goslings on the water, and even chickens on the banks, may be gulped by immense bullfrogs. One large frog of the Solomon Islands is recorded as catching birds, and the poisonous horned frogs of South America—already noted (Fig. 10)—catch small

Fig. 10.—Horned frog of South America (*Ceratophrys cornuta*).

mammals. Large frogs may sometimes turn the tables on the snakes and swallow the smaller ones. A snake eighteen inches long has been found in a frog's stomach. Fish and reptiles are sometimes eaten. In confinement frogs eat each other.

It is surprising what long fasts frogs are capable of, if the numerous records are to be trusted. There is no doubt that at certain seasons while they have not yet gone into the winter sleep, some frogs cease to eat—at least to any great extent—perhaps altogether. Of course when torpid in winter they do not eat. Dr. Abbott records that he kept a cricket-frog without food for one hundred days. At the end of seventy-five days it had lost only forty-four grains in weight. The author kept a common tree-toad in a bottle properly ventilated, one winter. It was always active when roused, but it could never be induced to eat. It finally died after many weeks, from what cause could not be seen, but no loss of flesh seemed evident.

In this connection it is proper to notice the wonderful stories we hear about amphibians being found in the hearts of trees, crevices of rocks, etc.—places not having any opening large enough, at the time of discovery, for the creature to crawl through. It would seem that it had been there a long time, and the query is double, How did it get there and what has it lived on ? Scientists are not much inclined to believe that such things have happened. But it is always best to see upon what such assertions are based. It is recorded that a frog has lived a year enclosed in a plaster cavity ; and Semper—a great naturalist—notes a definite instance of this kind where a Triton (a tailed form) was found enclosed in a cavity of rock from which an opening of one-twenty-fifth of an inch only in diameter and one-sixth of an

inch in depth ran to the outer world.   In this case
he thinks that the very young Triton crawled in
through this hole—not so large as a big broom-straw
—and grew so that it could not get out.   It was a
year or two old and two inches long when found, and
the naturalist thinks that sufficient food may have
strayed in there to support it.   It is probable that in
some other cases an outer opening may have been
overlooked.   At any rate the subject is an interesting
one in this connection.

It has been said that amphibians do not drink.
Just how this is proved in all cases the author does
not know.   It is well known that the frogs can condense water into their bodies by means of their skins,
or absorb it from green leaves by means of special
glands ; that they have a reservoir of pure water
within the body that is filled quite likely in this way.
To say that the aquatic kinds never drink is a broad
assertion, but that they may have no need to do so
may be true, because in this case also the skin may
merely absorb a sufficient quantity.

CHAPTER IV

CALLS AND MUSIC

IN those warm days in February when, in our middle latitude, the little male frog first awakes from his winter sleep and puts his head forth, the first cry is not for bread but for company to cheer his lonely heart; and he never ceases the croak or squeak till he finds it—or, at least, knows that the season is past for finding it, and that bachelorhood for another year stares him in the face.

Our ponds in the spring are thus made noisy by toads as well as frogs. In fact many of the early trills —especially those which are so prolonged—are from the toads (Fig. 11). Those who have had experience can tell what species is singing, as others can know the songs of birds. Dr. Abbott says that the little cricket-frog cries "pee-ceet" repeatedly; Dr. Jordan notes that the swamp tree-toad's call is like "the scraping of a coarse-toothed comb," and Professor Cope says the same is "a rattle with a rising inflection at

19

the end." The cry of our common tree-toad (14) is described as "a clear loud-trilled rattle." The

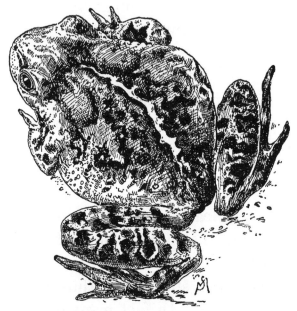

FIG. 11.—Common toad.

common green frog (Fig. 12) is called *Rana clamata* ("screaming frog") because it has a sort of barytone

FIG. 12.—Green frog (*Rana clamata*).

which it uses very frequently, and has the habit alse of "squeaking out" as it leaps into the water when

FIG. 13.—The bullfrog (*Rana catesbiana*).

disturbed. The spadefoot toad croaks fearfully in a deep rasp—as if his vocal apparatus needed oiling; and the voice of the bullfrog (Fig. 13)— especially when quiet and reflective in the later season — is described as " jug-er-rum," with a deep really musical ring at times. Again it may be a series of very explosive " chee-nngs " very far apart; but

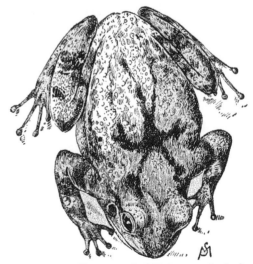

FIG. 14.—Tree-toad (*Hyla versicolor*).

in the spring when in concert with others it takes on various forms of squawks, croaks, etc., which may "make night hideous."

Frogs evidently sing in concert—even different kinds chiming in with others, as any one may observe; and among the bullfrogs there are leaders of the groups of singers, which seem to start first and thus get the whole band into pulsations or throbs of sound. In some tropical regions there is really a musical harmony in these concerts, and they are rather pleasant to human ears. It is known that some birds sing in harmony — the one making a good "second" to the other. But in many other cases the noise of frogs in the Tropics is so great that nervous persons sometimes have to leave the region.

As a rule amphibians do not cry out in fright, though our spring-frog (or green frog) is noted as an exception. Rarely do they express any sound in distress. A few groan under suffering, but usually they go silently to their doom, even while being swallowed alive by a snake. Neither do they as a rule express rage by sound. Dr. Lydekker, however, notes that the horned frog of South America (Fig. 10)—the one that has been noted as vicious and poisonous—defies its foes with a sort of bark, but that it has a clear bell-like tone for its friends. Perhaps then here low down in the backboned folk the language of rage is first separated from the language of love. Some of the fishes, however, had been calling for mates before this, for the heart is older than the head, and music older than speech. Perhaps the cat-fish when bellowing is defying his foe.

We shall not go into the structure of the larynx or special sound-apparatus in the frog-forms—further than to say that many of the males have membranous sacs on each side of the mouth which can be filled with air; and these greatly aid in producing a loud sound. In some they remain full so long as the pro-longed sound prevails; in others the sacs go down with the short call and are refilled before the next. The females do not have these sacs, but many of them call in a weak voice.

In the tailed forms there are calls also, especially from the land-haunters, but they are not strong nor striking. They are doubtless related to calling and charming.

Something of cries in connection with the weather will come up under " Skin."

WATER-HAUNTING

Among most amphibians that are land-haunters there is, in connection with the voice, the habit of forming bathing parties at the social seasons, when better opportunities of being agreeable to each other are afforded. Many of them hibernate at the bot-toms of shallow ponds and awake there; but others hibernate in holes on land and must awake, dig out and make this excursion to the water-party. It is an instinct in most animals below the mammals to at-tempt to rear their young in the place where they themselves were hatched. We see this very strikingly in birds and fishes. So the amphibians, being originally from the water, go back to it usually to bring up their babes.

In a general way water is necessary to the hatching of the amphibian egg and (since the young are usually fishlike), to the rearing of the tadpoles. There are a few exceptions to this now, though there was doubtless a time when the whole class resorted to water to lay eggs. In some cases, as on our dry plains, frogs and salamanders depend upon little temporary rain-pools in which to rear their young; so that here is one reason why amphibians should rejoice at the prospect of a storm.

OTHER CHARMING FEATURES

Having discussed the voice to call with and the place to assemble at, let us notice other means of charming or securing a mate among the amphibians.

Frogs are especially active in making themselves agreeable, though they do not resort to all the ceremonies of the "best society." But besides music and a decent bath the amphibians seem to condense our bowing, dancing, and posing into some very extravagant antics at times; and on such occasions they sport all the finery they can afford.

Many of them in every-day life are exposed to great dangers and must dress so as not to be seen easily, as is the case with the toad and many which are dull-colored; but others that live among pretty things can match their dresses (and their complexions) to their surroundings, and yet remain pretty.

Thus our wood-frog's coat looks as if it were crossed with twigs and plant stems and blotched with moss and leaves; the tree-toads are greenish or gray like leaves or lichens; and yet the pattern of these colors is pretty. In some of these, their under parts only may be beautifully orange or golden—often marbled, etc.; and as they swim above their mates these beauty spots may be ravishingly displayed.

Still others, which are active or can escape their foes in other ways, are gorgeously brilliant on the back and upper sides of the legs. This does not always expose them, for in the Tropics the tree-toads are said to be colored like the blossoms and fruits on the trees. Some of the salamanders, which can escape into the water or otherwise hide, are brightly striped—even rivaling the snakes in green and gold. Among many there is quite a tendency to be spotted on the sides in the regular " polka-dotted " way.

In dangerous creatures which are not liable to be attacked, there is often great brilliancy—perhaps because beauty is always desirable, and they can afford it. But a great naturalist has supposed that this beauty is a warning to the enemy—a warning which is, however, purely for the warner's benefit. Thus in South America there is a little frog that is conspicuously colored, but it has a very acrid skin-secretion, which keeps ducks and other things from eating it. It hops abroad fearlessly in daylight, and flaunts its gaudy colors defiantly. The horned frog (see Fig. 10) (not horned toad) is also brilliant with green and gold, and it fights and poisons.

COLOR CHANGES

To return to color as a protection, many amphibians have the ability to change their colors *at will* to suit the surrounding—to dress for the emergency. The common tree-toads all have it, and some terrestrial frogs also.

One of our little tree-toads is nearly solid green above, sometimes slightly spotted, but it is rare. The only specimen the author ever saw was on a green leaf, and the toad was solid green with no spots noticeable.

At the season when amphibians desire a mate, both sexes put on their brightest colors, and the males are not so noticeably the more brilliant here as they are in the birds and fishes.

OTHER ORNAMENTS

We have noticed, under their respective heads, the putting on of extra claws, webs, fins, and the enlargement of fingers and arms during the social time. In this connection it is noticeable that many males, like the birds, put special ornamentation and color on these also. The English male newt or eft (Fig. 15) has his vertical fin on the back much larger at this time than is the female's; and he has the edges of it all beautifully scalloped, as are sometimes the edges of collars, kerchiefs, etc., with the ladies. The end of the tail is similarly scalloped. He seems to think that his mate can appreciate beauty of form also. In this fin there are no muscles to move it, and in later months it disappears. It was an ornament only—a pretty thing—made out of an old implement that was once useful otherwise.

Fig. 15.—Crested newt (*Triton cristatus*). Lower figure, male; upper figure, female.

Sometimes the Axolotls (Figs. 16 and 17)—a Mexican tailed form—are found albino or white—from causes not understood.

Weapons

Besides the horns noted in the South American poisonous frogs (and these may be mere ornaments) the modern amphibians are scarcely endowed with special weapons—aside from tongue and teeth for prey-taking. Frogs have been recorded as fighting desperately with each other, some having had their bodies ripped open; but with what kind of weapons it is not stated. The teeth on the margin of the jaws may come into play. The spur-toed frogs noted as having claws on hind feet can probably scratch severely.

Salamanders when teased turn themselves suddenly like caterpillars and snap their jaws at the disturbing object. In general there is no such viciousness found in the amphibians as prevails in the reptiles, where the enemy is sometimes pursued and a bulldog kind of grip is often taken.

This sudden bending of the body is a favorite means of leaping by some tailed forms.

Skin-Secretions

As means of defense skin-secretions prevail more largely in amphibians than elsewhere, though some low mammals, as the opossums, possess them. In no case are such secretions agreeable to an enemy, but snakes do not seem to care for that of the toad, though it is nauseating to a dog. That of the brilliant little South American frog has been mentioned. The skin-secretion in the salamanders is very great. The ancients thought that it could resist fire—perhaps because this quantity of ooze might protect it a little. They thought the secretion deadly and blasting—even to vegetation. They imagined that it produced all sorts of spells even at long range. We now know that this is a fallacy, but its slime is really poisonous to lizards and small things which may get it in the mouth. So there may be a grain of truth in many myths. The impression prevails among many persons that the secretion of toads produces warts. Its own back is pointed out as a proof. But the wartlike lumps there are merely the glands from which the secretion flows. On the neck behind the eye these glands are larger and the fluid from these parts is usually more acrid.

A threatening attitude is a frequent means of escape or defense among amphibians. Toads normally have the sections of the breast-bones overlapping, so that they can swell themselves enormously when angry. Frogs have their breast-bone pieces meeting edge to edge, which prevents this power of expansion.

CHAPTER V

EGGS, SPAWNING PLACES, VIVIPAROUS FORMS, AND
PECULIAR CARE OF YOUNG IN AMPHIBIANS

Eggs

THE eggs of frogs and toads are, in general, like those of the higher fishes. They consist of dark dots that are yolks enclosed in a mass of jellylike matter, which is the "white." This "white" may be all in one great sheet or string; but in tailed forms, the eggs may be separate, buttonlike masses of "white" —each with a yolk in the center. In the upland frogs and toads, some of which do not lay their eggs in the water, the eggs may be separate, and placed singly here and there in crevices. These single eggs are apt to be much larger than those which are laid in masses. In the land-haunting tailed amphibians, the eggs are laid in packets or flat bunches. There is no shell or very tough membrane around the eggs of the amphibians, as in the birds and reptiles or sharklike fishes. In the cæcilians the eggs are also separated (Fig. 16).

The place where eggs of most amphibians are laid is in the water, generally a shallow stagnant pool. Usually they sink to the bottom or are twined around the stems of plants. Among the newts the mother

climbs the stem of a submerged plant and puts an egg
on each stem or leaf as she goes—one for each leaf;
but the axolotl swims over and among the plants and
may put more than one of her flat, buttonlike eggs,

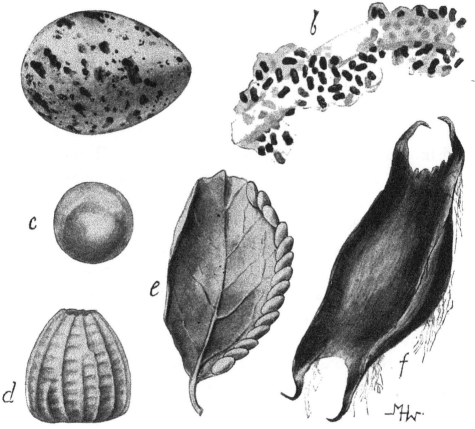

FIG. 16.—Eggs of different animals, showing variety in external
appearance. *a*, egg of bird; *b*, eggs of toad; *c*, egg of fish; *d*, egg
of butterfly; *e*, eggs of katydid on leaf; *f*, egg-case of skate.

all in a row, upon the stem. The Congo snake (*Am-
phiuma*) lays its eggs in a string in the water, then
coils its long body about them and bunches them into
a circular mass. So also the spotted triton (or spotted
salamander) bunches its eggs. The more land-haunt-

ing tailed forms lay their packets under damp moss or stones, etc. ; and here their young are hatched, and in some cases they never go to the water. They begin at once a terrestrial life, though the gills, which they have at first, show their aquatic origin.

TADPOLES

Generally speaking, the eggs of amphibians (Fig. 17) are hatched by the sun's heat. At first the little amphibian in its egg shows as a bloody streak, and appears to develop in its early stages much as a little fish. Later, however, it absorbs all the yolk into the stomach and does not have it suspended below the body as have the little fishes. From the egg the tadpole breaks away to liberty, if in the water, as a small plump beanlike body with a round sucking mouth and a slim wiggling tail. At first no gills are seen, but soon they grow as branches outside of the neck. Later these are lost, a hole is formed in the neck for breathing by gills that are developed *inside*—as in the fishes—a pretty strong hint that the amphibian did not get its gills from the present kind of fishes. This condition prevails only where the creature is going to be a land-haunter to any extent, as in the frog-forms and the salamanders. If it be destined to remain in the water, as in the case of sirens, mud-puppies ("waterdogs"), etc., these outside gills remain, and no inside ones are formed. Finally in the adult frog-forms and in the more upland tailed forms the holes on the side of the neck close, and the creatures become lungbreathers only. While no amphibian may wholly

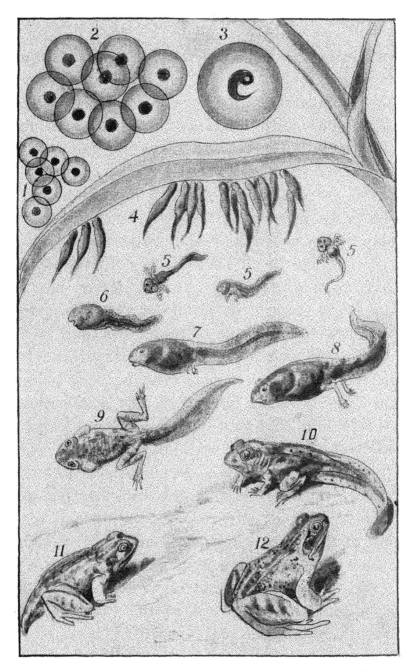

Fig. 17.—Development of the frog. 1, eggs when first laid; 2, eggs at a later stage; 3, egg containing embryo; 4, newly hatched larvæ or tadpoles; 5, 5, 5, tadpoles with external gills; 6 to 11, later stages in the development of the tadpole; 12, perfect frog.

reject the use of the lung, yet in the more aquatic kinds it has much degenerated or never developed, and these creatures die when out of the water as quickly as many fishes. Since these show a tendency to have weak limbs they are likely degenerate forms that have lost their high estate by laziness.

In all tadpoles which develop into land-haunters the limbs begin to appear about the time the lungs develop. It is remarkable that these limbs should be almost complete in all their parts before they come forth from under the skin, where for a time they form a small stump. It is one of those "shortened up" processes of Nature of which we shall find so many from this on. In the tadpoles of the frog-forms, modern demands have reached back so far as to grow out the hind limbs first, but in all others it is the fore limb that first shows. In frog-forms the tail soon departs—is not lost—but is absorbed into the body—as the limbs grow. In all others it remains, and has vertebræ (joints of the back-bone) form in it. There is never any vertebræ in the tail of the tadpole of the frogs, which have only horny jaws, and are rather more vegetable-eating than the other forms. It is said, however, that they have teeth before they have the beaks. In all tadpoles the gape is small— the mouth rather sucking. Those of toads, like the eggs, are always much blacker than those of frogs. It is impossible here to outline any further means of recognizing the various kinds of tadpoles. But an expert naturalist will know many if not all of them by some peculiarity. Those of frog-forms show only two pairs of gills on each side; those of some tailed forms (tritons, newts, etc.) have three pairs, while the shape of the body, tail, limbs, and the number of these last and their toes are all different in different species. The body in the tadpoles of frogs and toads is much shorter than that of the tailed forms, and in the former only there are sucking disks under the head.

The cæcilian tadpole shows a swimming tail, some internal gills only, and in one kind the stump of a leg —all of which are lost later when the creature begins to burrow.

Viviparous Amphibians

In many amphibians, as in fishes and reptiles, the eggs are hatched in the body before they are laid. In some instances the eggs are laid but are hatched immediately. In a few cases the entire tadpole-state is run within the body of the parent, and the young are born in the complete form. A remarkable instance is that of the Alpine salamander. Many eggs are formed in the mother but two only are hatched. All the others then run together in a mass to feed the two growing tadpoles. These have very long bushy gills which they lose at birth. They have been taken from the mother before they were born and put into water, whereupon they lost their first gills at once, but grew other new shorter ones—those by which they breathed in the parent's body, being quite likely too large for the much better aerated water. This, Mr. Mivart, a great naturalist, has cited as an instance where *at once* a creature could adapt itself to its surrounding, while young and plastic, without waiting for generations of the survival of those best fitted to it by mere accident.

Another species nearly akin, the nototrema, or pouched frog, hatches *all* of its many eggs within the body (Fig. 18).

It is said that in some kinds of cæcilians the young are born alive in the water, while another species certainly lays rather large eggs in a burrow near the water, and the mother coils above them and

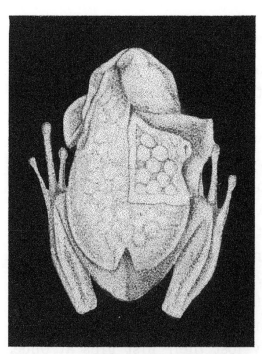

hatches them out, like a hen. If so this is probably the first instance of real incubation in nature. Some others which must have water to hatch their eggs, resort to queer methods to get it. One little West Indies tree-toad (Hylodes) lays her eggs at the point where the leaf joins the trunk in palms or similar trees. Here little pockets of water are found

FIG. 18.—Pouched frog (*Nototrema marsupiatum*). The brood-pouch opened to show the eggs.

after rains. Another of the same group takes the matter more by faith still. At the time when she lays there is usually a drought. So she places her eggs on limbs of trees above dried-up pools. Here they dry up also and are preserved, and when the rains finally come they are washed off and hatch in the pool below. Other species deposit them on the bottoms of dried pools.

Some toads have learned how to get along without water at any time. In the island of Guadeloupe, where marshes are not found, a little toad

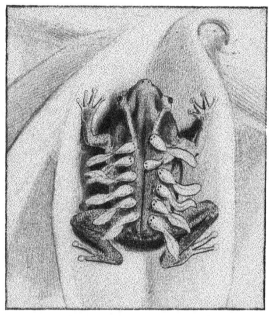

Fig. 19.—Tree-frog of Dutch Guiana (*Hylodes liniatus*), with tadpoles attached to her back. They do not fall off even when she leaps rapidly away.

places its eggs under damp leaves, and the whole tadpole-state is run within the egg, and the young come forth perfect.

There are various stages of taking care of the tadpoles when they form, without allowing them to remain in the water. It is well known that the female Surinam toad has a pitted skin at the breeding season, and that the male takes up the eggs and with his fore paws presses them into these pits. Here they swell, after the female enters the water, till each fills its cell, when a covering grows over them and remains till they

hatch and come out again perfect frogs, having, however, been tadpoles a little while in the pits. Several
other frogs have sacs on the back in which the eggs
are hatched. A tree-frog of Dutch Guiana, and also
one of Trinidad, carry their tadpoles around on their
backs, where the tadpoles cling by their peculiar sucking disks (Fig. 19).

The males of a great many frogs have a peculiar
habit. They take the strings of eggs and wind them
about their thighs. Some of these then go at once
into the water, while others sit in a burrow till the
eggs are ready to hatch and then go. Our spadefoot

FIG. 20.—Axolotl (*Amblystoma tigrinum*), tadpole state.

toads are said sometimes to do this. A frog in the
Solomon Islands, which we have noted as laying its

eggs in crevices, has its young hatched perfect—active and leaping.

The Mexican axolotl (Figs. 20 and 21) shows a peculiar form of suspended growth. If all the con-

Fig. 21.—Axolotl (*Amblystoma tigrinum*), adult.

ditions be not fair, it will cease to grow, and spend the rest of its life in the tadpole state, reproducing its young in this immature state while in the water, and these young are capable of either becoming land-haunters, by losing their gills, or remaining always aquatic. It is probable that other blunt-nosed salamanders do the same, or they may make the change the second year and not the first.

In the obstetric frog (Fig. 22), which wraps the string of eggs about his legs, the tadpoles are hatched in water, but they have no gills.

There is a number of nest builders among the

5

frogs. A Japanese frog makes a nest in the ground. Another in Brazil makes circular nests in shallow water, smoothing and shaping rings or atolls of mud, and laying its eggs in these cup-like depressions.

Fig. 22.—Obstetric frog (*Alytes obstetricans*), with strings of eggs.

CHAPTER VI

RESPIRATION

SINCE amphibians have no ribs to expand their lungs, those which breathe air get their breath by literally swallowing it, but they have muscles which expel it. In the water-haunters with gills, the lungs are mere sacs, without cells or pouches. In the cæcilians the left lung is small and nearly useless. All the kinds which stay under the water long have certain places in the body where well aerated blood is stored; and a large blood supply runs to the skin also. The skin aids the amphibians in breathing, even where there are good lungs, as in the frogs. To be thus useful it must be moist, like the gills of a fish, so that frogs and toads especially have water condensed into the body which they can cause to flow out over the skin.

In the tailed forms, the body has a "lateral line" or series of pores along the side of the body, like that of fishes, whence a secretion keeps the skin moist and slick. They have many other mucus glands besides.

We can thus see that frogs may breathe better

41

in damp weather, and hence the tree-toads rejoice at the prospect of rain.

Circulation

Amphibians, as a rule, are above fishes in that they have better hearts.

Although the heart is usually three-chambered, the blood is piped away from it in such a way that only a portion of it passes through the lungs or gills, the remainder going the round of the body again without reaching any aerating surface. They are not, therefore, warm-blooded. Before the lungs of tadpoles are used, the heart has only two chambers, as in fishes, while the blood runs from the heart through a pipe for each of the three gills on each side; but when the frog is grown, two of these tubes go to each lung and one other is absorbed. This is a noticeable step upward, since the warm-blooded creatures have only one of all these six tubes left, while the earliest fishes had eight.

In the frog which may sit part of the time with his rear parts in the water, and his foreparts in the air, there is a beautiful arrangement of pipes, valves and obstructing glands whereby Nature seems to compromise with him, and make the part of him in the air warmer-blooded than that in the water. Only in the crocodile (a reptile) elsewhere is there any such arrangement, and that is not just like this.

Lymph Circulation

The body of all creatures has a special fluid for carrying material for repair to the muscles, etc., and for bringing away the waste. This is called lymph. Nature always supplies a surplus of this liquid food ; and yet not being wasteful, she carries this lymph back and again pours it into the blood. To get this back there is in the higher animals a very large system of vessels, along which the fluid is pressed by the action of small vessels—as a sponge absorbs water. But in the amphibians and some other low forms, this fluid may move in large spaces between muscles or in long sacs (sinuses) under the skin or other membranes ; and since these easily expand under pressure the spongelike action (capillarity) does not move the fluid properly. So Nature has made these spaces and sacs (sinuses) to pulsate and thus send their contents onward. They are therefore called "lymph-hearts."

Now if we look closely at a frog, we can see places on its sides "beat" as if he had the "heaves" or "thumps," to use a horseman's words. There may be one or more fluttering places on each side, and they do not all throb at once or with any regularity with each other. There is one on each side of the tail. If you did not know how a frog breathed you might think that this pulsing was his way of getting his breath. In the amphibians these great lymph-canals often surround the blood-vessels ; but this is not the case in man. There are two of these lymph-hearts in some birds also, as the goose, at the root of the tail. Unlike other hearts they degenerate as the creature gets higher.

CHAPTER VII

SKIN, SMELL, HEARING, EYES, DIGESTIVE TRACT

Skin Shedding

THE skin of amphibians is shed frequently—sometimes at regular intervals; but the frequency depends upon many conditions of growth, health, etc. In the frog-forms some shed it once a week with great regularity, at certain seasons in summer. Again this may become quite irregular. In these, the skin tends to come off all in one piece, but there are instances where it is torn off in strips. Toads appear to get rather excited at this disrobing, and while the process may be usually easy there are times when the skin comes with great difficulty. They seem to call upon that internal reservoir of water to moisten the dried skin occasionally; and whether it come in strips or as a seamless whole, they invariably swallow it —sometimes rolling the mass into a ball with the hands.

The dry-land forms are said to shed their skins in strips, and these too are eaten at once. We should remember that this is not really the skin proper that is shed, but a thin, membranous—almost horny—outside covering called the *epidermis*. In all creatures

44

this must be got rid of in some way because it does not grow as the true skin beneath does. In all above the reptiles it is shed in little fragments, dropping off all the time or going with some special bath.

The true skin of the amphibians stays and enlarges with the body, as in other vertebrates. It is this that has in it the glands for secretions, the arteries for breathing, and which lies above the lymph-cavities, etc.—a great and important organ in every vertebrate.

SMELL

Amphibians are better endowed for smelling than are the fishes.

In the tadpoles the nostrils are mere depressions in the snout, not connected with the mouth, and they are then like those of most fishes. But in adult forms the nostrils open into the mouth, whereby the creature both breathes and smells by the air. The positions of these openings differ in the frog-forms and in the tailed forms. They differ in separate species of each group also, and are sometimes used in classification or description. There is much in the arrangement of the mucous membrane of the frog's nose which implies that it smells well. If the strong odors from the glands of the neck are used as charming perfumes (they are more likely for defense or defiance) this would hint that there must be fairly good smelling powers. But it does not take much nose to smell some odors—especially that of garlic, which the excretion of the toad resembles.

HEARING

All amphibians when adult have ears, but the tailed forms and some frogs have very poor ones which are devoid of any drum cavity. In most of them, however, there is an internal ear of some sort opening into the mouth—sometimes by one hole in the roof—sometimes by two, always behind those of the nostrils. The number and position of these holes aid in describing groups. In the lowest forms and in the tadpoles, the ear is a mere sac in a cavity of bone. In this sac a sort of chalky body (or bodies) is found as is the case in all higher ears. The cavity is simply covered with skin. No amphibian has an outside opening to the ear, but the higher frogs and toads have a drum cavity and a tough drum-membrane ôver it, which is flush with the surface. The size and shape of this membrane is very distinctive. In the genus *Rana* (bullfrog, green-frog, wood-frog, etc.) it is set in a sort of gristly ring and is very noticeable. Sometimes the drumhead seems itself to be a gristly plate.

There can be no doubt about frogs hearing well. While writing this book, the author stepped out to listen to some frogs in a pond one-fourth mile away, but he unfortunately let the door slam a little too hard, whereat the concert ceased. Similar experiences occur in trying to creep upon them.

There is much doubt, however, about their hearing high-pitched tones, or distinguishing changes. A frog "changes his tune" very slightly; and while certain students claim that there is some evidence of that part of the ear (*cochlea*) which appreciates pitch being found in frogs, it is, if there at all, very rudimentary. This may be the reason why they croak in such rasping quavers. Any one near a croaking frog can feel his ear-drums fairly flutter in the coarse vibrations.

As in fishes, it is not improbable that the sense-glands of the lateral line also may aid the tailed forms to appreciate jarring sounds.

EYES

In the frog-forms the eyes are very good, being usually fairly large and projecting. The eye is not so large anywhere as we should expect in creatures so nocturnal as many amphibians are, but this is probably accounted for in the great range of the size of the pupil. Those which, when examined in daylight, appear as slits are doubtless large and circular at night, as are those of the cat. Whether the pupil be a horizontal or a vertical slit, whether triangular or circular in daylight, these are very characteristic marks of various species. It seems that the vertical pupil implies more nocturnal habits than any other shape.

Frogs have some muscles which aid them in projecting the eye upward for observation. Those which haunt the water have projecting eyes which, with the nostrils, can be thrust above the water while the body is beneath. There is no partition of bone between the eye-sockets and the mouth, so that if the mouth be inflated the eyes project more. All the frog-forms have eyelids. The lower one tends to be transparent (like glass), and hence it has been said that they have a third lid or "nictitating membrane." There are no tear glands. Immersed in the water, the amphibian has no need for tears to wet the eye. Many frogs, like some fishes, can roll the balls over in the sockets and thus moisten them. The lids are moist from other sources. Frogs floating on the water are often seen to immerse the head suddenly and roll the eyes backward as if to wet them.

In the tailed forms the eyes are much smaller and less perfect. Some have eyelids, but in those which always keep their gills, either external or internal, the eye is usually much like that of some fishes, having no lids, but the outer skin runs directly over them. In *Proteus*, which lives in a cave in Austria, in *Typlotriton*, found in a cave in Missouri, and in the burrowing cæcilians the eyes are covered by the *thick* skin proper, and they remain as mere dots. While the lens is gone, enough of the nerve-matter of the eye remains to enable the creature to tell light from darkness. In one of the cæcilians this sort of eye has even sunk beneath the bones of the skull; but its tadpoles have better eyes, along with a fan tail, and a hint of a leg—all of which show how low these crawling creatures have fallen.

The flying tree-toad only has large, owl-like eyes, and needs to see at a distance, to make its tremendous leaps. In our slow-going common toad, which is also nocturnal, the eyes are small and dull.

Digestive Tract, etc.

In the grown-up amphibians, all of which avoid vegetable food in the wild state—at least very largely —the digestive tract is as simple as that of the fishes, more so than that of some fishes. In many, the stomach tends to be a mere swelled place in the long tube, and there are very few kinks or bends anywhere. There are no salivary glands, as in reptiles, and the other organs, as liver, kidneys, etc., occur, but their uses are much simpler than in the higher creatures.

The tadpoles, however, are so largely vegetable feeders that the digestive tract is long and much twisted, as it always is where tough matter is to be digested. But while the gills are being lost, the limbs growing, and the lungs forming, it shortens up into a much simpler form and takes a sudden step backward—an instance of another wonderful emergency met almost in a moment.

CHAPTER VIII

SKELETON GENERALLY—BACK-BONE, RIBS, SKULL, MUS-
CLES, NERVES, REFLEX ACTION, TENACITY OF LIFE,
AND REPAIR IN AMPHIBIANS

SKELETON

THE skeleton of the amphibians is interesting for both what it has and what it has not (Fig. 23). The back-bone in the lower forms is much like that in the lower fishes. In some fossil forms the original gristly string around which the back-bone is built still remains. In many others the ends of the vertebræ (or pieces of the back-bone) are flattish, or merely a little cupped at both ends—a very primitive state, like that of the sharks.

In the frog-forms, however, are found the most interesting peculiarities of skeleton. The number of vertebræ are very few—those of the tail being gone as noted; and instead of many joints in the rear part of the body there is one long, unjointed rod, which runs from about the middle of the back to the rear end of the body.

Note that the rear legs are attached far back, near the point where the tail should be, and not well up on the back-bone as they are even in man, and that

behind the junction there are only two little vertebræ
to represent the tail bones. Note also that the forward end of this bony rod (called the *urostyle*) has on
each side a projection against which the bones to which
the legs are fastened join directly. This really makes
three stiff rods side by side in the back here to resist

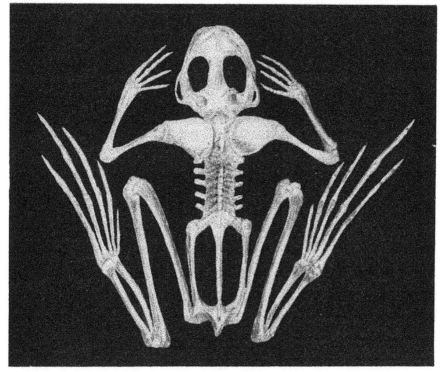

Fig. 23.—Skeleton of frog.

the sudden thrust of the powerful hind legs. Many
creatures leap, but none have themselves hurled forward by a stroke directly at the rear end of the spinal
column, as the frogs. These projecting pieces at the
forward end of the rod are called " the transverse
processes of the sacral vertebræ," and may or may
not be more or less expanded at their outer ends,

by which peculiarity frog-forms may be arranged into groups or classified.

You can see that there are no ribs—just mere "transverse processes." Only one family of frogs has any hint of true ribs. In the skeleton of salamanders there are pieces of ribs which in cæcilians are longer.

If we turn the skeleton of our frog over, we shall find that while there are no ribs to meet it, there is a very respectable breast-bone (*sternum*) to which the fore limbs are well anchored. This is the first real breast-bone in Nature, though the fishes have hinted at it. The tailed amphibians have it in gristle only.

SKULL

A noticeable amount of open space is seen in the top view of the skull of our frog. Opposite is an enlarged figure of the head (Fig. 24). Much of it in life is gristly, and in the lowest tailed forms it is so much more so that Professor Huxley has said that here it is little better than that of the lamprey—a low form of the fishes. In the long-ago, however, the monsters of the class had more bones and more *bone* in the roof of the head, as you may see from Figs. 31 and 32, page 63.

NERVOUS SYSTEM

The amphibians — especially the frogs — show many peculiarities of the nervous system not found so strikingly in the mammals. They retain in their bodies hints of their low ancestry ; and indications toward a more intelligent condition are found in their heads. We may glance at this briefly.

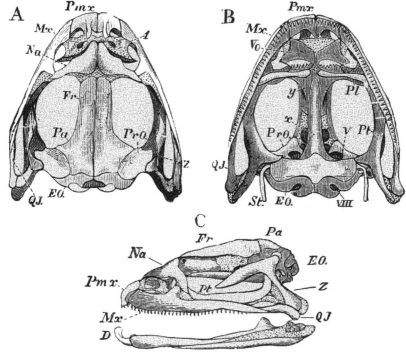

FIG. 24.—Skull of *Rana esculenta.* A, from above ; B, from below ; C, from the left side. *x*, parasphenoid ; *y*, girdle-bone ; *Z*, the "temporo-mastoid."

The lowest animal tissue, where no nerves are apparent, seems to be able to feel, or to draw back or go on when touched. In this case it is supposed that feeling goes through any part in any direction, from one cell to another, where there is more than one cell. But as in time there came to be special cells for digesting, breathing, etc., so there came to be special cells for feeling and for stimulating other cells into action. These arranged themselves in rows, so they could communicate with each other, and these rows were the beginnings of nerves. They can be recognized as having distinct form and structure in those creatures which are well up the scale but yet far below the back-bone. In time the two duties of nerves, that of feeling things and stimulating muscles into action, were separated also, and separate rows of nerve-cells were given to each duty, though the lines lay close alongside. At the inner end of these two nerve-threads there was a union, which swelled into a little knot called a ganglion. It was simply a little crude brain, which received the news from the outer edge of the creature by one thread, and sent word back by the other, telling the members out there what to do. It was merely the rebounding place where the sensation returned and became stimulation. For a long time these little brains lay disconnected from one another as the early nerve-cells did. They were the lords of their own little realms. Each small margin or filament of the low creatures had its little brain to report to and to obey, and literally the right side (there were no hands then) knew not what the left was doing.

But in time these little brains became connected by nerve-threads, or else they massed themselves into bunches; and soon these *bunches* took control of larger areas of tissue; but we can not attempt to follow this development, which doubtless continued till a great confederacy was formed—which was massed in the back-bone—and then much later a seat of government arose at the forward end of this, which we call the brain. But *all* of these little brains did not come into the mass. Many were left out for a kind of picket work, and they still acted according to their own will, to some extent, and did not always telegraph to headquarters (literally) for instructions or orders. Nor can the central government always control them wholly. Thus, if the sole of your foot is tickled you jerk it away, and your brain or will has little to do with it; often can not prevent your foot from jerking. This is because there is a lot of little brains away down there which have very small connection with the big brain proper; and before the latter can have anything to say, the sensation has come to these, and from them the stimulation has gone back—reflected—to the foot. Hence this reflection is called "*reflex action.*"

In the movements of the heart and other inside organs, and in the opening and closing of the pupil of the eye, etc., the action is more independent still in answer to outside stimulation.

The amphibians have a large arrangement of nerve-matter for this purely reflex action—this kind of unconscious work which is not controlled by the brain.

If a frog's head be cut off the body will still be able to move and perform a great many acts which seem intelligent though it will never move of its own will. Something outside of itself must stimulate it. If the skin be pricked here it will scratch the place on this side; if there, it uses the foot on the other side. If that foot be cut off it uses the opposite foot and stretches it across the body. If the body be

6

turned over it will right itself. In some of the low creatures (as starfish) if a limb be cut off and laid upside down, this lone limb will right itself by the nerve matter in it. All this is the so-called reflex action. Perhaps in the frog the stimulation goes to the spinal cord, but it can not in the starfish.

Now if we hurt both sides of the frog at once, but make one side more painful than the other, the headless creature moves away from the worst pain. One reflex action—the stronger—overcomes another, and what appears to be an intelligent act may come in as the combined result of many merely reflex actions. Thus we may see how with a proper arrangement of these, all under the guidance of one great ganglion— even one so inferior as the frog's spinal column—intelligence or mind may arise in a certain form.

The intelligence of the amphibians is not remarkable, but toads and even salamanders become quite tame, and, in their indolent way, make interesting pets, coming to be fed at a call or whistle.

We have turned aside here to this little outline of the nervous system because nowhere else in the vertebrates are there so many interesting peculiarities all in one group.

REPAIR

It is because, partly, of this peculiar nervous arrangement that amphibians can so readily repair injuries or renew lost parts; and doubtless for the same reason their sufferings are not so great under wounds as are those of more conscious or less automatic beings. Below the amphibians, below the fishes, there are many creatures which not only grow new parts, but can grow new individuals out of each old part; because their nerve-matter is so arranged that no serious separations are made by the cutting. In more advanced creatures—even in the high fishes—the concentration of the great nerve-centers is too complete for the best repair work. A leg of an axolotl will be reproduced in a month, and a tail, replacing one that is lost, will soon grow out again with new vertebræ or bones forming in it. It is said that the bones do not again form in the regrown tails of lizards—only gristle. Fishes often eat off the gills of water-newts, and these are readily regrown. But in all cases of legs and tails, though they may be regrown repeatedly, the new ones are rarely so perfect as the old ones.

Hibernation

All forms of amphibians hibernate in winter.

Some dive into the mud at the bottoms of pools, some dig burrows, some crawl into crevices. Methods differ in species close akin. Some terrestrial tailed forms hibernate on land. On the other hand, some bury themselves in mud in summer and sleep away months in the tropics—awaking again with the rainy season, as is the case with some fishes (the lung-fishes) which are in many respects quite like amphibians.

The common toad, the cæcilians, the spadefoot toad, the obstetric frogs, and some salamanders burrow and spend much time at any season in holes. Some of the tailed forms are known to revive after being frozen solid.

CHAPTER IX

FOSSILS, KEYS, ETC.

Amphibians of the Far Past

Except the cæcilians, living amphibians are far away from the old fossil forms which had such peculiar teeth (usually), strong armors, and bony skulls. Perhaps they became so stiff and awkward that they could not escape from their enemies; or they may have become so inflexible in their structure that they could not change, as the conditions of the air and earth changed, and hence they perished.

The cæcilians, as noted, have something of scales and style of skull which was formerly fashionable. Their burrowing habits may have saved them, and the very humble habits of our little denizens of the slime may have preserved them also. Many connecting links between these and modern forms have perished. There seems to be no form yet found that stands between the frog-forms and the tailed forms of to-day; nor between either and the cæcilians. No salamander leaps much; no frog has a vestige of a tail outside of the body when grown. In their babyhood only the two groups come close together. As far back as frogs are found in the rocks, they are all *frogs* and their tadpoles can be recognized even as differing from others. The fossils of all living amphibians are in rocks that are modern compared with those in which rest the monsters of the class. Hence there has been plenty of time for the fossil forms to degenerate into the kinds now living.

FIG. 25.—Slab of sandstone with amphibian footprints, from coal-measures of Pennsylvania, × ¹/₅.

But no fossil among those giants approached the form of a frog. No reptile or amphibian which can be recognized as such is found fossil away down where fishes are so abundant. Just under the coal period some amphibians show, and just above it some reptiles.

According to Professor Le Conte the first traces
of an amphibian ever found were some tracks in an
ancient mud-flat near Pottsville, Pa. It was the footprints of one of the giant labyrinthodonts, breaking

FIG. 26.—Jaw of *Dendrerpeton acadeanum*, and section of tooth,
enlarged. (After DAWSON.)

into the records as a creature with fully developed
limbs, whose ancestry had lived long enough to lose
one finger, as you may see by the cut (Fig. 25).
Then the next was found in a stump which was set,
petrified also, into a great table of rock (Fig. 27).
This one was quite reptilian in structure. Above is a
cut of its jaw (Fig. 26).   Figure 28 shows the *Orche-
gosaurus* which is quite fishlike.   It was three

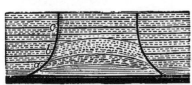

FIG. 27.—Section of hollow
Sigillaria stump filled with
sandstone. (After DAWSON.)

and one-half feet long. It
was *Ganoid* in scales and
had both lungs and gills as
some *Ganoid* fishes yet have;
and it is about the best
known connecting link be-
tween the old monsters, the
sturgeonlike fishes and the living amphibians. Recently hosts of little labyrinthodonts have been found
in Ohio. They had sharp noses and snakelike, limbless bodies (Fig. 29). There were some of these old

FIG. 28.—*Archegosaurus.*

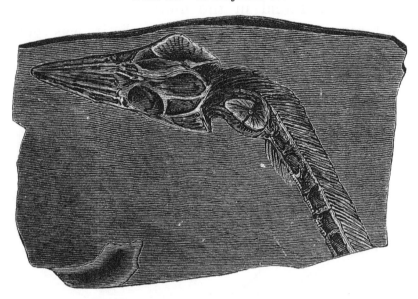

FIG. 29.—*Ptyonius.* (After COPE.)

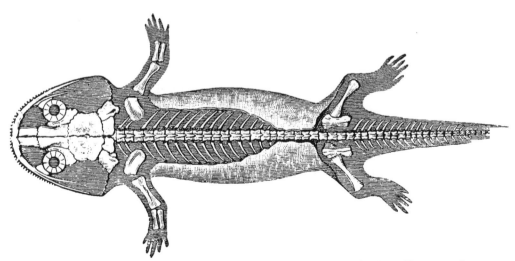

FIG. 30.—*Limnerpeton laticeps,* natural size. (After FRITSCH.)

forms, it is said, which did not have the skull so completely roofed with bone. Fig. 30 is also a cut of a small salamanderlike form which is, however, still a labyrinthodont, when its teeth are examined. Note that it has slight ribs, and that the skull-roof is complete. It is found in the more modern upper carboniferous, and it looks as though it was getting

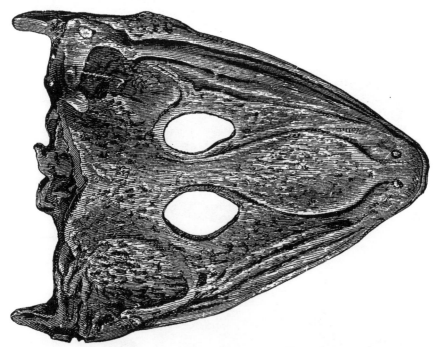

FIG. 31.—*Mastodonsaurus Jœgeri.*

near to the living kinds. But higher still and more recent (in the Triassic) there lived a monster with a head two feet wide and three feet long (see Fig. 31). It was called *Mastodonsaurus*, and Fig. 32 is a cut of the head and jaws of *Trematosaurus*—though neither were really saurians. But saurians (lizardlike reptiles) and amphibians had not got so far

apart then as now. Fig. 33 is a cut of the true labyrinthodont tooth already noted.

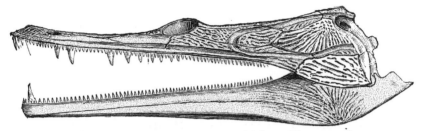

FIG. 32.—*Trematosaurus*. (After HUXLEY.)

In the Triassic age the frogs appear, and we won-
der what it was that made the amphibians lose their
wrinkled teeth, set in the bones of the jaw, and
allowed the reptile only to bring these on up to the
present time.

Professor Huxley remarks that since amphibians
seem to possess characters which belong to each of
the groups of vertebrates known as
Ganoids, shark-forms and lung-
fishes; and since these are
known to be well sepa-
rated from each other
very far below the
place where any fos-
sils of amphibians
are found, it is
quite probable
that these latter
branched from
the parent back-

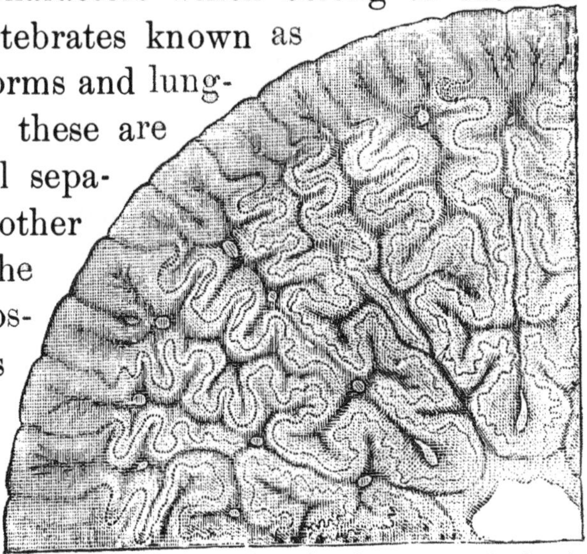

FIG. 33.—Section of tooth of a labyrinthodont.

boned ancestor at about the same time that the others
did, and hence are very ancient. To the author it

seems highly probable that this is true, and that the reptiles, which we shall discuss next, branched off independently at about the same time or at least very low down on the amphibian stem. Hence both are very ancient.

CLASSIFICATION

This class of vertebrate animals is characterized by two states of existence—one aquatic the other terrestrial—at least by a larval form in the egg or out of it. Eggs are always formed and true limbs are always indicated at some stage of life. The eggs are small, the body (now) uncovered, and the skull joins the back-bone by *two* ball-and-socket joints as in the mammals; but the lower jaw is hinged to the skull by special bones, which is not the case in mammals. In these last only the lower jaw is hinged directly to the floor of the skull. The following is a key to the orders of the amphibians, both fossil and living; then follows a key to the tailed forms, and finally there is a key to our common frog-forms in the Eastern United States.

ORDERS OF AMPHIBIANS

NOTE.—If the specimen is not described at the single letter, say A, go on to where the letter is doubled, as AA.

A. Skull roofed with bone—at least behind the eye. Teeth often wrinkled. Fossil. STEGOCEPHALA.

AA. Skull not so roofed.

  B. Legs absent. CÆCILIANS.

  BB. Legs present two or four.

    C. Tail present. *Salamander-forms.*

    CC. Tail absent. *Frog-forms.*

      D. Tongue present.

        E. Gristles of breast-bones overlapping.

          Suborder TOADS.

        EE. Gristle of breast-bone not overlapping.

          Suborder FROGS.

      DD. Tongue absent.

        Suborder *Tongueless Frogs.*

The salamander-forms are divided into families by various modes of classifications, based on anatomical differences; but the following simple artificial key will, by outside features purely, lead to the families (as now divided):

TAILED FORMS

A. *Outside gills gone in the adult.*

  B. Eyelids present, no gill-opening (the real salamander-forms).

    C. Tail round—no fin. *Salamanders.*

    CC. Tail flat—with a fin above. *Newts.*

  BB. Eyelids absent.

    D. Toes two or three behind and three in front.

      Congo Snake.

    DD. Toes five behind, four in front.

      Giant Salamander.

AA. *Outside gills present in adult.*

  E. Limbs four. (Proteus), *Mud-puppies.*

  EE. Limbs two. *Sirens.*

The frog-forms have their suborders scientifically divided into many families, genera, and species. The discussion is too great for our space and too technical for our plan.

To know such as the reader is apt to meet in the Northeastern United States the following may be helpful:

A. Teeth absent from upper jaw.
    B. Skin warty, toes webbed.   *Common Toad.*
    BB. Skin smooth, toes free.   (*Toothless Frogs?*)
AA. Teeth present in upper jaw.
    C. *Fingers and toes with slight dilatations or pads at tips.*
          (Hylidæ, our) TREE-TOADS.
      D. Webs absent on fingers; pads mere dots.
        E. Brownish above; head green.   *"Cricket Frog."*
        EE. Grayish above; no green or brownish.
          Swamp Tree-toad.
      DD. Webs present on fingers; pads large, shotlike.
        E. Greenish above.
          G. The green has a yellowish or olive cast; some spots on back (as well as the sides).
          Hyla squirrella.
          GG. The green pure—pea green; no spots on back.
          Green Tree-toad, Hyla andersonii.
        EE. Not greenish above; yellowish drab or dusk-colored.
        Pickering's Tree-toad, Hyla pickeringii.
CC. *Fingers and toes not dilated or padded at tips;* they end in sharp points.   (*Rana.*)
    H. Spots on the back squarish, their edges or outlines nearly straight.
      I. Back greenish; spots not in straight rows; thighs with three broad bars.   *Leopard Frog.*
      II. Back brownish; spots rectangular, in rows; those on thigh not forming broad bars.   *Pickerel Frog.*
      HH. Spots on back *not* squarish; either round dots or irregular blotches.
        K. Web of feet not reaching the tip of the fourth toe.   *Green Frog* or *Spring Frog.*
        KK. Web of feet reaching tip of fourth toe.
          Bullfrog.

Other frogs are found in our region, but they are not so common as these.

PART II

STORY OF THE REPTILES

By JAMES NEWTON BASKETT, M. A.

STORY OF THE REPTILES

CHAPTER X

INTRODUCTION, DEFINITION; WHAT CAME IN WITH THE REPTILES; ORDERS, LIMBS, TOES, CLAWS, TOE-WALKING

THE Reptiles are known from Amphibians, as we have seen, by their scaly bodies, and by having no gills at any time, and also by having the head joined to the neck by only *one* ball-and-socket joint instead of by two. The tongues also of the two classes differ. Nearly all reptiles and some fishes are scaly, but the scales of the two classes are usually very different. Those of most fishes, when present, can be scraped off, or are loose and outside of the skin; while those of the reptiles are mere horny folds of the skin itself and do not come away. A few reptiles and many fishes are scaleless, however, but no reptile has gills or gill-openings, while no fish is without both of these; they are thus distinguished the one from the other.

As we go upward, the rule is, scales for reptiles, feathers for birds, and hair for mammals. If we had lived in one of the long-ago geological periods (Jurassic or lower), we should doubtless have seen creatures half-bird half-reptile; and feathers and scales would have been mixed all over the body on at least one creature, as we find them now mixed on the legs of birds. Then also a little lower, perhaps, the mammals and reptiles could not have been distinguished from each other by their covering—or, indeed, by anything else; for all classes were very much merged into each other at an early date. Even now the pangolins, the armadillos, and other mammals show scales and plates; so that some reference to internal anatomy is necessary in certain cases to distinguish reptiles and mammals.

Many distinguishing features might be mentioned, but the presence of glands for nourishing the young by milk is peculiar to no class but mammals. It gives them their name, thus separating them from all others. Outwardly, then, a reptile may be defined as a strictly lung-breathing, cold-blooded vertebrate usually covered with scales or horny plates, while the young are hatched from large eggs and are never nourished by means of milk-glands, and never have a tadpole state.

Besides the complete abandonment of gill-breathing there is found now with the reptiles the first sternum or breast-bone having the ribs completely reaching it. As noted, there is some evidence that amphibians once had ribs nearly complete, but have lost them. The fishes hinted at the breast-bone, but it was useless; the amphibians had it to swing the fore limbs to but not to join the ribs to, but in the reptiles it first becomes an implement of respiration, whereby the lungs are made to open and shut. In tortoiselike reptiles it is absent, and the ribs are stiffened into the shell, but a muscle called the diaphragm

—the muscle inside of us which hiccoughs—helps to force the breath out; and the reptiles have the honor of introducing this muscle also. In serpents there is no breast-bone, but a great array of long ribs, that almost encircle the body, help them to breathe. The reptiles also, through the crocodiles only, brought in the first four-chambered heart, and hinted first of hot blood. Thus have all the creatures shown their progress by their breathing and circulation. We shall see that some other things came in first with the reptiles, but we shall note them later. This little pre· view is given that we may know why we should be interested in this class—a class which in its backward ties and upward outlook has no equal.

THE GREAT GROUPS OR "ORDERS"

As we glance at the living (not extinct) reptiles they seem, like the amphibians, to be divided by their forms into three great groups: First, the tortoise-forms; second, the lizard-forms; and third, the serpent-forms. But this will not hold with the scientist —except in the case of the tortoise-forms. He tells us that the crocodiles, though lizardlike in shape are far from being so in structure, and really a much older family; that another lizard-shaped creature in New Zealand (*Sphenodon*, *Tuatera* or *Hatteria*) actually belongs to one of the old families further back still, and that there is considerable doubt whether lizards and serpents should be separated at all, since some snakes have rudiments of legs and some lizards have none at all. He would even hint that the chameleon should be separated from the lizards. We have seen that outward form is not a safe guide, since a lizard and a salamander may have the same general shape without being nearly related.

FIG. 34.—A serpent.

For our purposes we shall speak of the reptiles under the following orders, and we shall learn their peculiarities later: tortoises, serpents, lizards, crocodiles, and tuateras. The last three have legs and a tail like those of lizards.

FIG. 35.—A lizard.

The tortoises have shells over the body; the crocodilians have plates placed edge to edge; the lizards and serpents have overlapping scales; and on the

FIG. 36.—Crocodiles.

tuatera the skin is warty. The serpents are practically legless. We know enough now to begin to learn some-

FIG. 37.—Tuatera.

thing further. Besides these there were once many forms, now extinct, the peculiarities of some of which will be referred to as we go along.

LIMBS

The limbs of the reptiles are rather like those of the amphibians in a general way, except that the claws are well developed. The webs of the toes are not so noticeable, though the tortoises, crocodiles, and tuatera have swimming membranes. All reptiles swim well, however, and the tortoise-forms, crocodiles, and some snakes are especially aquatic. Many fingers and toes, rather than few, prevail generally, though there are some remarkable exceptions. The number of toes may run from one to five. Normally there are five before and five behind, but where the limbs tend to be lost, the toes decrease also, till in a certain skink-like lizard there is only one toe behind, and in some greaved lizards there is only one finger in front.

In the sea-tortoises the toes are all massed into paddles which are often much like fins, except that they have the three divisions of the leg, a characteristic of all quadrupeds; and in some ancient forms (*Mosasaurs* and *Ichthyosaurs*) the limbs were still more fused and flattened. In the fossil *Ichthyosaurs*, some species were found which had six, possibly seven, rows of bones inside the paddles. It seems probable, in one case, that the two outside rows were merely extra bones on each side of the original five fingers for they are not joined to the hard bones properly; but in another case both feet had six good toes. If this state of affairs had continued on down, or up, to man, we should not be counting now by tens or decimals, but by twelves or duodecimals—a really much more convenient system if we were only used to it; for while ten has only two factors, twelve has four.

Whether these old swimming reptiles had gained these toes extra or inherited them from the fishes, and whether the others have lost all but five, can not be determined. If they were once land-haunters and went back to the water, Nature may have spread the foot for them, as she has the paddle of the whale, by putting in extra bones. If they came of ancestors which were always aquatic — having acquired their good lungs and good three-jointed limbs while yet in or near the water, as the amphibians did, then the five-toed land-haunting animals have lost a sixth toe. It is said that there is a hint of this in some frogs. Against

FIG. 38.—Foot of a chameleon, showing how the toes are bunched together, and opposed to each other, in grasping an object.

this last view lies the fact that a fin or flipper does not need to be three-jointed to be used as such, while a good walking limb certainly does — which facts argue slightly for a land origin for all three-jointed limbs, whether legs or paddles.

As a rule, there is not any marked opposition of the thumb or big toe in the reptiles. In the chameleon proper (not our little Florida lizard, so-called) the toes are bunched wonderfully (for grasping) into twos on one side of a twig and threes opposite (Fig. 38).

But most reptiles with limbs climb by claws, or claws and toe-pads combined, as in the geckos (see Fig. 60). One order of fossil reptiles had the little finger greatly lengthened, by which it doubtless flew by means of a skin-membrane attached.

CLAWS

The claw, as such, came in fully with the toed reptiles, and is now often sharp and clinging. In the tortoises claws are present to aid the creature in scrambling along, and in burrowing. In fact these creatures walk almost exclusively by the claws, or push by them rather. But the more aquatic turtles have some missing usually. The pond-turtle omits one; and those with flippers may have only two on each limb. In the crocodilians, where the toes are four behind and five in front, there are only three claws to each foot.

Another thing which came in with the reptiles more fully was the act of walking on the toes only, leaving the heel high up. This is a practice found in many mammals, such as dogs, horses, etc. Most reptiles are flat-footed walkers, however, while some others, such as the frilled lizard, like the mammalian raccoon, are flat-footed when going slowly and toe-walkers when in a hurry.

The number of joints in the toes of lizards is especially interesting in that they have the same order in number that occurs in the birds. The bird, however, lacks the fifth toe. The first toe has two joints; the second toe, three joints; the third toe, four joints; the fourth toe, five joints; and the fifth toe the same as the third, four joints. But in some old paddle-limbed kinds of reptiles there were a great many joints in the digits, as there are in the paddles of the fringe-finned and other fishes now.

Besides the serpents, many lizards are limbless, as are the *Amphisbœna* (no English name), and the so-called slow worms (*Anguis*) of the Old World, the glass-snake or joint-snake (Fig. 39) (which is a lizard —*Ophisaurus*) of America, and many others found in the families of skinks, greaved lizards, and other groups. In some of these the rear pair of limbs only may remain, and in others the fore pair only are present.

It is well known that most serpents are limbless; but the family of crushing or constricting snakes (*Boidœ*), boas, pythons, and anacondas, and many of their near-by kin, show rudiments or stumps of limbs at the rear end of the body. In some other families near to these, the stumps do not show, yet the little bones to which the hind legs are usually attached— the so-called pelvic girdle—are found beneath the skin. But no vestige of a fore limb, or of the "shoulder girdle" even, is ever found in a serpent; and no lizard—though appearing legless in front or everywhere—has ever been found without these shoulder-bones. Hence, by dissection a limbless liz-

FIG. 39.—Glass-snake (*Opheosaurus ventralis*). The tail is twice the length of the body, and breaks off at the slightest blow. When broken off it grows again.

ard may be known from a snake; but we shall see later that the tongue also will usually distinguish all limbless forms.

By the walking on the toes only, the reptiles brought in the first outlook for speed afoot which finds such high development in the running birds and the strictly toe-walking mammals, such as the horse, the antelope, the greyhound, etc.

In keeping with this, one of the extinct reptiles had its toes hoofed instead of clawed.

CHAPTER XI

TAILS

In the reptiles the tail seems quite important, for no reptile, except the *Amphisbœnœ*, is without one; and even in this family some even show stumps. These creatures run backward, and a tail would be in the way here. In some sea-turtles it is very short, as it is in some of the dry-land kinds. These latter, when they close their shells, take great pains to get the tail well boxed in.

In some extinct lizard-forms, known generally as *Dinosaurs*, the tail acted as a fifth limb or prop as they walked, stood, or sat erect on the two hind legs only; and these tails·must have been terrible weapons, as that of the crocodilian is yet.

Our smaller lizards retain the tail for various uses, and doubtless for ornament also. Some of the large monitors can strike serious blows with it. In others, as the chameleon, flying lizard, and some tree-lizards, it is prehensile and can be curled around a limb to aid them in clinging and climbing. Doubtless the tail in lizards, as in the salamanders, is a means of expressing the emotions, and, since we find it (alone) highly colored occasionally, it is probably an ornament also.   Some run with it curled over the back like a scorpion's, and such lizards have been wrongly called "scorpions" because of this habit.   In this connection, some lizards have a peculiar use of the tail which is found in other creatures as well, but not frequently.   It is that of making with it a sort of unconscious prayer and sacrifice for the safety of the body.   In the European lizards, in our glass- or jointed-snakes (see Fig. 39), and others, not only are the bones of the tail loosely attached to each other, but they have a sort of membrane, which runs between the joints and extends outward through the muscles and skin even.   By this means the whole tail is "jointed" and the parts may be separated, without loss of much blood, as the parts of an orange come apart, without any loss of the juice.

If a pursuing enemy grasp this tail, it breaks off readily and may allow the body to escape, as if the creature thought it better to go maimed into salvation than to go whole into destruction—especially since the part lost, in such cases, is soon regrown.

Seriously there is no *thinking* about it, by the creature.   In some instances the exertion even of trying to escape may break off the tail of our glass-snake, and leave it wriggling for a while to attract the enemy's attention; and so purely mechanical is this action that sometimes the body itself has been known to turn and swallow the squirming thing. All stories about these parts reassembling are myths.

The new tail simply grows again, and no part of the body breaks.

On the contrary, one lizard has a tail set with spines all around nearly as numerous as hairs (Fig. 40), and if this be left outside when escaping into a burrow it is not a savory mouthful to the pursuer.

FIG. 40.—Spine-tailed lizard (*Uromastic spinipes*) and young.

The ancient forms of lizards often had great spines on their tails which were very effective weapons. Others had the tail flexible and flat for swimming purposes, as it now is in crocodilians.

In snakes the tails taper with the body usually, and thus complete the symmetry or beautiful shape, but they are useful in many other respects. In the

sea-snakes (Fig. 41) they are flat and fringed like those
of eels, and they are thus the means of swimming.
In the land-snakes, tails are helpful in springing and
running.   Our " spreadhead " (*Heterodon*) sometimes
makes great leaps down-hill by this means, and our

FIG. 41.—Sea-snake (*Hydrophis cyanocincta*).

common " blue-racer " (blacksnake) can erect its body
half its length and run rather rapidly on what must
be mainly the tail.   All the tree-haunting and the
constricting, or crushing snakes, use the tail to cling
with, and to aid them in climbing and anchoring
themselves while crushing or holding their prey.
The whip-snakes (Fig. 42) and other tree-snakes have
tails that are longer than the body, wherewith they

FIG. 42.—Whip- or tree-snake (*Passarita myeterizans*).

tie themselves almost as if with a string while they hurl their remaining length almost as a bolt upon their prey below them.

In some burrowing snakes the tails are very short and blunt; and in one family—the shield-tails (Fig. 43)—there is a shieldlike button at the end which better enables them to push the body through the

FIG. 43.—Shield-tail snake (*Silybura macrolepus*). These remarkable snakes look as though their tails had been cut short off. In some species the body ends in a naked disk, in others with a rough horny point, in others again, as in the species illustrated, the disk is covered with keeled scales.

earth. Others have a sort of horny tip for the same purpose, as our common pine snake, and in some cases in the Old World kinds there are broad scales beneath with sharp, backward-set spines on them which are helpful in pushing the creature along or in.

This horny shield reminds us that rattlesnakes have a series of horny rings upon the end of the tail, by vibrating which a buzzing sound is made that is a warning or threat of anger or attack. Some harmless snakes rapidly vibrate the tail against a dead leaf or other object and thus produce a similar sound for similar purposes—perhaps an imitation. Many others vibrate the tail, but not necessarily against anything.

Among the extinct flying reptiles, the kind known strictly as the *Pterodactyls* had no more tail than the

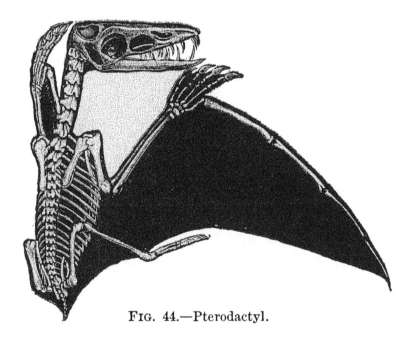

FIG. 44.—Pterodactyl.

modern birds (Fig. 44), but another kind, called the *Rhamphorynchus*, had a most preposterous racket-shaped affair, like that of a windmill, which must have been used as the tail of a kite to hold the creature against the wind, in which direction only could they probably fly (Fig. 45).

Heads

The heads of the tortoise-forms are, in a rude way, quite birdlike, ending as they do in a toothless horny beak which often has on it a downward hook at the tip. That of tuatera (or *Hatteria*) is more turtlelike than most others which are not turtles, though certain

FIG. 45.—Rhamphorhynchus.

lizards tend to have horny beaks. In all these, however, there are teeth. In a general way the heads of snakes and lizards are much alike, though in some snakes the neck is very much smaller than the head, and the latter is then apt to be diamond-shaped. In fact not till she got to the reptiles did Nature seem much concerned about the neck, but at an early date among the fossils some of these necks were extremely long and flexible. The heads of crocodilians are long and flat, with a slight neck evident, which is smaller than either head or body. But in lizards and tortoises the neck is usually about as large as the head.

Perhaps in all modern reptiles the head extends

in the same line as the neck, as it does in nearly all fishes and amphibians; but in many extinct forms of reptiles the head was placed at right angles to the neck, as it is in the horse and so many other mammals. This doubtless resulted from the high elevation of the forepart of the body in these old monsters. All modern forms are primarily crawlers, and hence the low horizontal head and neck.

In perhaps all reptiles the size of the head is very small in proportion to that of the body; and in some fossil monsters it was so absurdly small as to make us feel that the creatures to which they belonged had just sense enough to feed themselves and to walk around.

The heads of crocodilians have the skin tightly drawn over the skull and the bones are much carved or sculptured. The skin here is not covered with horny plates or scales as it is in most lizard-forms. Some lizards have beneath the skin a shield of loose bones which are not a part of the skull or skeleton proper.

JAWS

The jaws of the reptiles are very interesting to the student. In all the vertebrates below the mammals, the upper jaw has some slight movement upon the skull, though it is in no sense hinged as the lower. This is especially true of the beaked kinds. In the crocodilians, the upper jaw appears hinged as they lie flat with the mouth open, but it is really the whole head that is lifted. It is true that the flat

8

head is well fitted for this, but if you will lay your lower jaw on the table and open your mouth the lower will not move but the whole upper jaw will lift the head up and back (Fig. 46). The lower jaws of reptiles are peculiar in that each side is made up

of a great many separate bones, usually, though not always, grown together. In the higher animals there are not so many. As noted, the reptiles and all below them hang the jaw to the skull by one or more bones— often by only one, the so-called "quadrate." Usually this is hinged or loose, as in snakes and most lizards; but in tortoise-forms,

FIG. 46.—A crocodile (*Crocodilus niloticus*) lying with its mouth open, showing the apparent movement of the upper jaw instead of the lower one.

tuatera, the crocodilians and the chameleon, it is fast to the skull in various ways, of which the classifier makes much. In the snakes the bone to which the *quadrate* hangs is itself loosely hung to the side of the skull, so that the jaw can be pried well away from the head as their bulky prey passes into the throat (Fig. 51). Here also the two halves of the jaw never fuse together in front, but are tied together merely by an elastic ligament which allows them to spread apart in swallowing large objects. Again, this and the double hinge at the skull allows the jaw to be thrust forward, first that on one side then on the other, so that the mouth is thus worked over the prey by the backward-curved teeth—one side holding what is gained while the other advances. The snake thus literally gets over (or " outside of ") its prey.

There was an old fossil monster called *Mosasaurus* (which was a lizard, but quite serpentlike) that had a better arrangement still. In the middle of each side of the lower jaw was a joint bending downward and outward. On the front part were backward-set teeth. Its jaws also were capable of moving first one side then the other; but you can see that every time it bit its prey the joints straightened like a nearly open jack-knife and pushed the front part forward by the pressure of the bite.

Something similar to this, though not just like it, is found now in the upper jaws of those poisonous serpents which have fangs that lie down when the mouth is closed but are erect when the mouth is open. By means of a joint in the middle of the upper jaw (which is pulled straight by the muscles as the mouth opens), the bone lying across the upper end of the mouth, to which the fangs are fastened, is rolled downward and forward, thus letting down the deadly fangs.

By means of the separation of the lower jaw at the chin, snakes are known from lizards, and it will be observed that there is a marked difference otherwise.

The jaws of mammals all have an upward projection upon the jaw itself, which is formed purposely to meet the skull, but in all other creatures the skull itself sends down the bony projection—either loose or securely set in such direction as to meet the jaw.

The jaws of serpents are rarely used for crushing or killing, but largely for seizing, holding, and slipping the throat over the food, and in the poisonous kinds, for forcing in the fangs. Snakes are strictly swallowers, and their whole head-skeleton is arranged for this practice.

Teeth

In the reptiles Nature seems to have experimented with all kinds of teeth. Here she seems to have made useful the wrinkled or grooved sorts found in the ganoid fishes and labyrinthodont amphibians. She made the grooves the channel for poisons, and even folded some of their edges in till they became tubular. But more of that later. While many lizards and all serpents, perhaps, have teeth somewhere on the roof of the mouth (to speak generally) it was in the reptiles that teeth first became confined to the jaws only; yet, in a few cases, Nature has made the most preposterons effort in this class by projecting the lower spines of the back-bone through into the swallow tube and putting enamel upon them, so that several species of serpents which eat eggs may have them broken after they are partially swallowed, thereby losing none of the liquid contents (Fig. 47).

Many of the fossil monsters were terribly armed with teeth that grew in sockets or grooves directly out of the jaw-bones. Some also had teeth set in several

FIG. 47.—Dasypeltis-unicolor, in act of swallowing a fowl's egg.

rows or pavements, which were used evidently to grind vegetable food, and some had beaklike, duck-shaped jaws, like those of the spoonbill. Others had rather turtlelike, or birdlike, beaks with a pair of great tusks projecting, and others had mouths armed with short sharp teeth, and in their midst were terrible fangs, like those of dogs and tigers (Figs. 48, 49, and 50).

These Professor Cope called "*Theromorphs*" or beast-forms, because their teeth were so very much

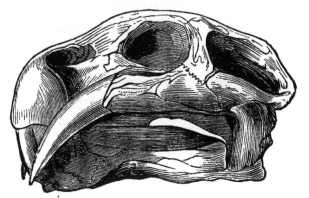

Fig. 48.—Dicynodon lacerticeps.

like those of some modern mammals (or beasts). The grinding teeth here first began also to show cusps or more points than one. *Læ-*

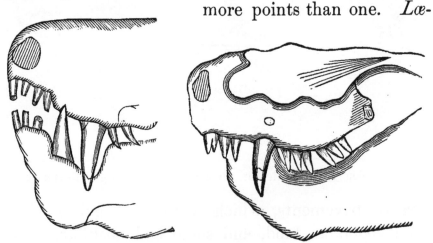

Fig. 49.—Lycosaurus.

laps, a terribly clawed carnivorous fossil reptile, had teeth that were serrate (saw-toothed).

In the crocodilians, the teeth grow much as in some of the old monsters. They come up out of the

jaw-bone and are renewed by one pushing out the
hollowed and partially absorbed tip of the other;
but they have many sets, and the new teeth below
the gums are said to be "nested" into each other as

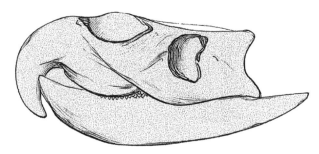

FIG. 50.—Rhynchosaurs-Hyperodapedon; Trias (after Huxley).

are thimbles. As a rule, modern lizards have their
teeth grown down to or up from the jaw-bone, though
they are not set in it, but are fast to it. In the
Tuatera there are two front peculiar teeth which are
a little like those of rodents (rats, rabbits), but which
fuse together and form almost a beak above.

In lizards some teeth are conical, some serrate, as
in *Iguana*, and some are flat and merely crushing or
grinding—according to food. Nearly all lizards renew their teeth by having the new one form directly
beneath the old; but in the *Anguidæ* (slow worm)
the new grow between the old.

The teeth of serpents are usually recurved, sharp
conical points.

The erectile poison-fangs mentioned are always
found with other smaller ones (to the number of three
or four) concealed beneath the flesh behind them,
which are thus ready to rise up and take the place

of the forward active one, should it be broken; so that jerking the poison-fangs out of a rattlesnake makes it harmless for a short time only. The new ones do not have to grow much, but merely *rise* into place (Fig. 51).

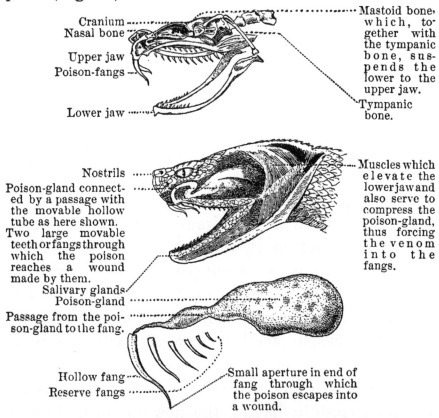

Cranium
Nasal bone
Upper jaw
Poison-fangs
Lower jaw

Mastoid bone, which, together with the tympanic bone, suspends the lower to the upper jaw.
Tympanic bone.

Nostrils
Poison-gland connected by a passage with the movable hollow tube as here shown. Two large movable teeth or fangs through which the poison reaches a wound made by them.
Salivary glands
Poison-gland
Passage from the poison-gland to the fang.

Muscles which elevate the lower jaw and also serve to compress the poison-gland, thus forcing the venom into the fangs.

Hollow fang
Reserve fangs

Small aperture in end of fang through which the poison escapes into a wound.

FIG. 51.—1, skull of rattlesnake, showing the manner in which the upper jaw is connected with the lower one; 2, head of rattlesnake dissected to show poison-glands, etc.; 3, poison-gland of rattlesnake.

TONGUES

The tongues of reptiles are various in shapes. So far as known none are tied down in front only, or are *largely* free behind, as in the amphibians. The front is free if it be free anywhere, and two points may project backward from the rear edge, as may be seen in the tongues of birds. These points are aids in swallowing. In one genus of lizards (*Chaleis*) these forks are especially long.

In the tortoise-forms, the tongue is usually short, flat, and cupped, rather fleshy and smooth, as if it were a tasting organ. It is much like that of some fishes, and, within small limits, very movable, though it can not be thrust out. In the crocodilians it is fast to the lower jaw all around and acts merely as a floor to the mouth.

In the lizards it takes on two extreme forms generally, with many shapes between these. In most it is flat and much the same thickness everywhere (forked behind or (not), and is usually notched in front. This kind of tongue rarely runs in any sheath throughout. One type of this form is flattish and runs in a sheath at the base only; and another sort is thick at the base, thin and wide at the tip, which latter runs under a sheath or strap.

The other form of tongue is long, slim, and deeply forked at the tip. Sometimes it consists of only two mere threads. This is the kind found in all snakes and two or more large families of lizards. In the snakes and many lizards (monitors, etc.) this slim tongue is entirely sheathed when inside the mouth, and is thrust forth very rapidly either for feeling or threatening; but the tongue itself is perfectly harmless—even for securing prey.

In other lizards this slim, forked tongue is covered with scales, or deep wrinkles, or rough, brushlike points (like the tongue of a cat), and it must, therefore, be used to grasp small objects or assist in chew-

FIG. 52.—Anolis or American chameleon (*Anolis principalis*). Although the general color of the animal beneath is white, the upper parts may quickly assume hues varying from a vivid emerald green to a dark iridescent bronze color.

ing them. This kind is noticeable in the greaved lizards. In a few others, the tongue is said to be spearlike at the tip, somewhat like those of wood peckers, and it is evidently a capturing implement.

In all the true lizards of the Old World the tongue is forked and smooth, but not sheathed. In the family of the skinks, which includes our blue-tailed and ground-lizards, the tongue is only slightly notched, and is rough or scaly; but in the family of the *Iguanidæ*, which includes our so-called "chameleon" (Fig. 52), the common little "swift lizard," and all the host of horned toads (*Phrynosoma*), the tongue is smooth, short, and barely notched, and it can be put out a slight distance only.

The chameleon proper has a tongue which it can expand at the end at will, and thrust far out by means of a long stretchy stem—thus easily capturing insects.

CHAPTER XII

FOOD

THE food of the Reptiles is various. The tortoise-
forms are largely flesh-eaters, catching fish, frogs,
floating water-birds—anything; but some are vegetable feeders, such as the green turtle, renowned for
soups, and the case is the same to some extent with
the sea-turtles; but the "hawkbill" and "logger-
head" and leather turtle are carnivorous. The crocodiles are known, of course, to be fearfully carnivorous
(flesh-eating). They may approach large prey near
the shore and strike it into the water with their tails
or grab it suddenly with their jaws and draw it under
water and drown it. They usually stow it away in
some cavern or safe place till it partially decays, when
they bring it to the surface later to eat it.

The chameleon's diet is one of insects especially,
and not even a frog is more highly equipped for their
capture.

So far as the author knows no snake is at all inclined to feed on anything vegetable, though many

98

eat worms and insects, and drink milk. In all cases snakes do not chew their food, but gulp it, often while it is yet alive. It is well known, however, that the great crushing snakes suffocate their prey before swallowing it, and that the poisonous kinds kill it first with their fangs, and then eat it some time after. Their poison causes the flesh to tend rapidly to decay and thus aids in digestion. All the reptiles, unlike the amphibians, have salivary glands, and in the serpents these are large. As snakes begin to swallow their prey these glands are very active, but the snakes do not slime their victims over with the tongue, as is often reported. They doubtless pass it over their victims for the purpose of examination, for the tongue is their best investigating organ. Perhaps the size of objects swallowed even by the anaconda has been much exaggerated. A sheep or a calf or other small young cattle is about the limit of what they can do in this respect.

We have noted the special arrangement of teeth in serpents. It is said that some tree-snakes—not poisonous—which capture birds, have an extra long tooth, designed perhaps for penetrating through the feathers.

As to a snake's ability to charm a bird there is much uncertainty and some strong assertions and denials. But it is certain, at least, that the presence of a snake is often so terrorizing to some small creatures that they seem unable to move or escape, and that birds do often approach a quiet serpent gradually nearer and nearer till they come within its reach.

It may be a sort of madness of attack, and not any special "charm." The author once witnessed a summer yellow-bird so behaving, but he was prevented from seeing the end by the noise of others approaching. The testimony of many concerning actual capture having taken place in this manner is sufficiently worthy of belief. But the *kind* of attraction or paralyzing effect exerted is by no means settled.

Lizards are both carnivorous and vegetable eaters. In a few cases, like the snakes, they eat each other, though there is not anywhere now a "lizard-of-prey" (corresponding to the bird-of-prey or preying mammals) which is adapted to devour its kind, as was the old *Lælaps*—a reptile of a past age.

Most lizards are fond of insects. Many found in the Western States eat leaves, buds, and blossoms of plants. One of these, the "chuck-walla," is a large, fat, lazy lizard, faring well on this weak diet. There is a sea-lizard that haunts rocks by the ocean and eats seaweed. Many of the giants among the fossils had peculiar methods of feeding, as we may infer from their teeth; but we will note these later when we mention the families. The land-monsters were mostly browsers, while the sea-monsters were carnivorous.

OFFENSE, DEFENSE, AND ESCAPE

Nowhere are there more various offensive and defensive methods, or means of being disagreeable, than appear in the class of the Reptiles. While all are not well endowed, some are armed and armored wondrously. We have already spoken of the teeth, which are weapons not only against prey but against enemies. The poison-fang and its sac or gland full of deadly fluid is the most terrible of close-range weapons. It is overcome only by means of superior strength, armor, or activity.

In the large fossil forms there were many weapons proper. Besides the terrible array of tusklike teeth, some *Dinosaurs* had special spurs on the paws, and others had their large tails armed with spines. Another still had many horns about the snout, and a spiked collar of immense spines about the neck, and others had these along the back. One, already noted (*Lælaps*), had long, curved, tearing talons on the rear feet, and walked erect, and was able to strike down prey much larger than itself. In no modern form do claws play a special part as weapons.

HORNS

But several lizards and some snakes have apparent horns, which may be weapons proper, not connected with prey-taking. Quite likely they are often useful in fighting or tantalizing a rival only. Among a few lizards, as our so-called horned toad (Fig. 53), battle consists in the turning of each other over on the back. It is rather more of a wrestle than a fight, and the one flipped topsy-turvy "gives up" at once. While in these "toads" there is no special horn or hook on the tip of the snout, yet in some other lizards these are present, as may be seen in Fig. 54. Wherever these are found, they are on the male often and not on the female. This happens frequently in the chameleons, where in one case there are as many as three horns. In another there is a peculiar forked prolongation of the snout.

Among the snakes some vipers have horns—sometimes one on the tip of the snout, sometimes two— one over each eye. Their use can scarcely be under-

FIG. 53.—The horned toad.

stood. It is said to be the rule that snakes do not fight as rivals; and it is fairly well known that the bite of a poisonous snake is not harmless to his brother, and often not injurious to other non-poisonous kinds. Dr. S. Weir Mitchell states that he has repeatedly injected the poison of snakes into their own bodies and seen no ill effects from it; and a correspondent of the author (a scientific collector) states that he has frequently boxed rattlesnakes and non-poisonous sorts together and observed them bite one another without ill effects. But more recently a Paris experimenter claims that one snake is affected by the venom of another in proportion as it is itself poisonous. This should cause innocent snakes to suffer. It is very certain and frequently observed that our black-snake and others which are non-venomous will attack and destroy the rattlesnake. The superiority in this case has usually been attributed to quickness and strength preventing the rattler from striking; but the safety may possibly consist in the scales acting as an armor. Certain it is that the poison-fang is not usually a weapon for fighting a rival, but it is said, however, that non-poisonous snakes die of the bite of sea-snakes. The question can scarcely be said to be settled yet.

DEFENSIVE ARMOR, SPINES, ETC.

The most striking armor now found in Nature is that of so-called box-turtles where every part of the body is protected. That of the armadillos is almost as good, however. The shells of all tortoise-forms are not so complete as these, and may consist (in a low form) of a mere cap over the body and a mere cross or strip of shell on the bottom. The soft-shelled or leathery sort of turtles have the outer covering above skinny, leathery, or gristly. But in all there is a layer of bones beneath which " breaks joints" in a rough way with the usual array of horny scutes outside. Likewise in many lizards there are flat bones on the back beneath the skin.

While no ancient reptile had just any such armor as this, we may see that many were rich in bony plates and spines which were very effective. A hint of this remains in the crocodilians. Here there are thick horny plates placed edge to edge, so strong that they

9

formerly turned balls of the old muzzle-loading musket.

Of course the scales of lizards and snakes are shields, but, as noticed, some lizards have bones in or beneath the skin over the back and rear part of the head. A few lizards, the chameleons, and one snake are merely warty or tough-skinned.

Besides this, many modern lizards (as well as the old fossils) are plentifully protected with spines.

FIG. 54.—Head of leguan (*Iguana rhinolophus*).

These may run along the back and upper edge of the tail only, as in the common iguana, *Tuatera*, crocodiles, the leguan (Fig. 54), the Galapagos sea-lizard, and others; or they may be all over the body, as in our horned toad of the Western plains and the Moloch lizard of Australia (Fig. 55). These spines all grow from scutes that are buried in the flesh or skin, but the tips of every scale are slightly spinous in some lizards.

TERRIFYING METHODS

Besides actually hurting their foes, many reptiles terrify or threaten when they are disturbed, and some have special bluffing implements. To some extent the spines of the horned toads are such, and the creature swells the body so as to make these spines more projecting and the body less easily swallowed.

Fig. 55.—Moloch lizard of Australia (*Moloch horridus*).

Others, however, swell their non-spinous bodies. In others still—especially the frilled lizard—there are frills, flaps, wattles, etc., which are erected threateningly—the one mentioned having a great frill around the neck, "like a Queen Elizabeth collar," which it turns forward over its head at its foe, and walking erect on two feet, with tail elevated and mouth wide open, it makes a terrifying dash at an enemy (Fig. 56).

We can not notice all of these peculiarities, but a

most striking example of what is perhaps pure bluff
is found in some species of the so-called "horned
toad." When irritated, they throw from the corner

FIG. 56.—Frilled lizard, standing at bay with frill erect. Running,
    showing similarity to running pheasant. Foot, showing how a
    three-toed track can be made with a five-toed foot. Running
    erect, posterior view.

of the eye a little jet of blood upon their disturber.
Members of the U. S. Biological Survey, and others,
have experimented with this curious habit, and have
found the fluid to be real blood, and that the jet can
be repeated from either eye.

It is hard to see the object of this, unless the
creature hopes to make the foe feel that it has
wounded him—perhaps by one of the spines—and
that he would better withdraw. Since the blood of
some animals injected into that of others is poisonous, it may be that it hopes to hit some place where
the skin is broken and thus poison its enemy. But
it is more probable that it is merely intended to
frighten. It is well known that grasshoppers and
other insects exude a harmless fluid when caught, and
that some caterpillars and beetles eject a very hurtful
acid at a disturber. It is well established that some
poisonous serpents can eject their venom many feet;
but this seems to be the result of their attempt to
bite at the enemy, whereby the fluid is squeezed out
of the hollow fangs. The same muscle that closes the
jaws compresses the poison sac.

ODOR

The Reptiles are, as a class, very bountifully sup-
plied with glands for secreting and pores for emitting
odorous fluids—not, however, for projecting to great
distances as in the skunks. While some of these
odors are for defense, most are likely for charming or
being agreeable, just as the beau or belle of to-day
uses the musk of an animal for the same purpose.

But in a few instances, as in the musk-turtle, some lizards, snakes, and others, it is used as a protection against enemies, and is very successful so far as the human foe is concerned. Odor is doubtless a means that reptiles have of advertising their position to each other at social times. The lizards, in some cases, are distinguished from each other by the presence or absence of pores on the thighs for emitting the odorous secretion. Crocodilians have similar pores under the throat as well. Some old fishermen have stated that the odor from these forward pores of the alligator is attractive to fishes, and the musky creature thereby gets a living by its perfume. This is not confirmed, but is not very improbable. During the battles of these monsters this odor can be detected miles away, down wind.

ORNAMENTS

We have been compelled to say that certain things that appear as weapons, etc., may be merely ornaments; for weapons are frequently ornamented in Nature. But there are among the reptiles many appendages which are ornaments purely. To be brief, these are mostly found in the males and consist of frills, wattles, dewlaps, or great hanging folds of skin, and even the spines, warts, and horns are ornamentally located. These, when the social season comes on, have much brighter colors than they have in the winter, and some of them are erected, inflated, or spread out by the proud possessor when his sweetheart or his rival comes around.

Color

As noted, color comes in as a charming feature. Often the males are much the brighter. In the tortoise-forms the sexes are alike, but both are often beautifully marked and tinted. We can see this in tortoise-shell. Crocodiles are not especially charming in this respect, but many lizards are gorgeously colored. This is apt to be the case with tree-haunting forms, just as it is likewise among the arboreal snakes also that the brilliant colors and remarkable patterns are found. We can not even name the instances in either group. The family known as the *Elapidæ* (genera *Elaps*)—of which the little coral snake of the Southern States is our only member, but which are very abundant and very poisonous in South America —take on in rings, spots, and blotches all the brilliant reds, yellows, etc., of the most dazzling ribbon.

Color-protection

In the tropics many of these brilliant colors are protective, because the strong greens of the tree-snakes may resemble leaves or grass, and the other brilliant hues resemble flowers and fruit. In perhaps a few instances the patterns may imitate something surrounding, as it is so well known to do in many insects and some birds. One snake is especially noted as having its scales colored in groups of fives which are strung along the back and resemble the petals of a flower; so that hanging in a tree it may appear as a festoon of blossoms.   Our own little green snakes are hard to see in the grass.

So likewise many small reptiles are sand-colored, or resemble the dried dirt, dead leaves, etc., where they hide.   Many snakes and lizards also have the ability to change their colors in the manner of some frogs and the well-known chameleon.   Our little Florida lizard (*Anolis*) (see Fig. 52) is as good at this as any of them, hence its spurious name.   In this case the colors are doubly protective.   In the chameleon proper a lot of little colored granules or cells lie far beneath the skin, and certain ones of these can be brought to the surface by special muscles, and others depressed.   This is to be done consciously by the creature, which seems to know what color is required—since it has been found that blind chameleons do not change colors, but remain at their darkest in all lights or on all hues.

CHAPTER XIII

MOTION, HIDING-PLACES, HIBERNATION, HAUNTS, DISTRI-
BUTION, MIGRATION, PLAY, BATTLE, ENEMIES, DIS-
EASE, AGE, AND SIZE OF REPTILES

MEANS OF MOTION

NOWHERE has Nature been more liberal in modes of motion than in the Reptiles. Here she has run the whole scale. Many swim; some wriggle only; others burrow; most walk on four feet, a few on two; one glides or sails on the air like a flying squirrel; while another ancient form doubtless had well-sustained flight, like that of a bat. As variations of these methods some have leaped on two legs as a kangaroo, and others have waded in a sort of upright, half-floating way in deep water. This record can not be excelled in any class of creatures. Besides mere wriggling, the snakes (having lost their limbs by indulging too largely in that) seem to have needed some means of slow, gradual motion; whereupon Nature loosed the hold of the ribs from the breast-bone, caused the bone to absorb, brought the tips of the ribs to the lowest surface, connected them with the scales below, and strung to each a separate active muscle. Along the back-bone she put ribs the entire length, and in the back-bone she put extra vertebræ (to the number of 300 in the pythons) till the body was long and active. Each of these joints she made of large balls and sockets, and on the spines of the vertebræ she made other unusually large joining (articulating) surfaces in addition to those already in the centers; so that the back-bone should not only be bent easily but stiffened quickly and surely for good work. It is wonderful to see what snakes can do with a back-bone only and a slight movement of ribs and scales. They climb, leap, swim, stand erect for half their length, and in a few cases run swiftly. The author has seen the common garter-snake resting head downward on the rough bark of a standing tree, the diameter of which was equal to the length of the serpent; and he has noted the common blacksnake, not four feet long, run, as fast as a man would walk, through standing timothy two feet high, holding its head well up above the grass.

The movement of the tortoise-forms is often merely a sliding one. Usually the breast is *pushed* along by the claws. This is the case with many if not all living lizards. But, while the slow progress of the tortoise is proverbial, some of them can run with the body clear of the ground and a few make considerable speed in a dash for safety.

Those turtles with paddles swim rapidly and capture fishes even by dashing at them or pursuing them.

The running of some lizards is so rapid as to produce the effect of a mere streak, but it is not long kept up. We have noted that one of the frilled lizards has become so strictly toe-walking as to make only three toe-tracks, the two outer toes being so much shorter than the others that they do not touch the earth when the foot is stretched up (Fig. 56).

The flying-dragon (Fig. 57)—one of the tree-lizards —has the most remarkable apparatus in Nature outside of the birds for gliding down on the air.

In the fossil flying lizard (Figs. 44 and 76) already noted, the flight was by a member attached to the little finger. In fact, bats and even birds fly by their fingers. Flying mammals glide on the air by means of a fold of skin stretched along the body which is attached to and spread out by the limbs. But this modern flying lizard spreads a similar membrane by means of *its ribs*, which project outside of the body, another most remarkable use of these bones. They can be folded down by the side when not in use. There is no power of fluttering, however.

Crocodilians are able to make quite vigorous dashes at an enemy on land, but since the projections on the sides of the back-bone are long and close together, they can not bend the neck or the body much, and are therefore unable to turn quickly. They may be dodged by a leap to one side. In the water they swim well by means mostly of the flattened tail, but they are said to roll over and over when they have caught an animal, that they may confuse and more quickly drown it.

Perhaps it may be worth while here to remark that reports of a snake's progress by means of taking its tail in the mouth and rolling as a hoop is a myth,

found in the minds mostly of Southern negroes. No
snake so rolls, and none has any such weapon on the
tail as the horn which is said always to be present
in such venomousness as to kill the tree in a few

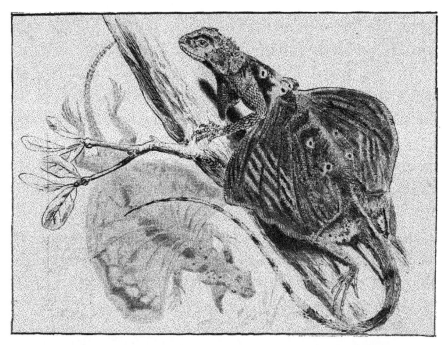

Fig. 57.—Flying lizards (*Draco volans*). They do not really fly as
birds do, but glide through the air like flying squirrels.

minutes which, according to the story, these snakes
always strike. No snake known really has any stinging or poisonous spear or horn on its tail—nothing
that is a weapon; though, as noted, there are points
and shields there which aid in progress. Our common
"spread-head" (*Heterodon*) does sometimes throw itself into great vertical loops when escaping down-hill,
notwithstanding the frequent statements from scientists that all snake-motion is a horizontal wriggle.

It may be possible that this occurrence is the basis of the myth just noted. Usually a snake's movements are horizontal undulations. All snakes swim in this manner, and the push of the finned and flattened tail of the sea-snakes is very effective.

Hiding-places

Perhaps as a way of escape hiding-places should not be omitted. Nearly all reptiles have something of the sort always near. Lizards run into crevices and climb trees. A few snakes burrow to escape immediate danger; others have holes of some other creature near by to slip into. The habit of the rattlesnakes of the plains in living with the social burrowing owl, in the homes of the prairie dogs, is so well known as to scarcely deserve mention. On our more Eastern prairies the gopher-holes are used by our short thick rattlers. Brush-heaps are favorite places for many innocent kinds. The land tortoises may burrow—those known in the South as "gophers" very deeply. Mud-turtles fall off of logs into water, and alligators drop all the body beneath the surface except the nose, or else float along safe in their protective resemblance to a half-rotted log. They are said also to have caves dug in the banks, where they hide.

Hibernation

Of course hibernation is another form of hiding to escape both the winter and an enemy. Where the cold is severe all reptiles may hibernate. The more aquatic tortoise-forms sink into the bottoms of ponds and streams; the more terrestrial burrow—some only to a very shallow depth. A pet tortoise may get very anxious about burying itself long before severe cold comes on. Snakes seek deep crevices in rocks, and all kinds, of different species, are said to ball themselves together—often in a mixed mass of various species. Lizards hibernate in any crevice. Crocodiles do not live generally far enough north to need a winter sleep. But in South America and other extremely hot haunts—where long droughts occur—these crocodilians bury themselves in the mud and lie incased in dried-up dirt, as is the case with some fishes and amphibians, till the rains come again.

HAUNTS AND DISTRIBUTION

Perhaps enough has been said to indicate the homes of the various reptiles. Their distribution over the earth is peculiar, and so extensive that it can not be discussed here. They are a so much older race than the mammals and birds that the same natural areas for these latter do not answer for the reptiles, which doubtless populated a large area before the others came in. Unlike the amphibians, they were not so much confined by salt water, though they seem to be largely fresh-water forms, now and in fossil times.

MIGRATION

Reptiles may be said to migrate only in the sense that some, as the sea-turtles, are known to resort year after year to the same island or sand-bank for the purpose of laying their eggs. When the conditions remain the same it is probable that all land-reptiles spend their lives very close to one spot. Snakes are noted in autumn as gathering toward their hibernating places.

Turtles, however, migrate from pond to pond, either on account of drought or to hunt new fields for feeding, etc. They soon find out a new pond, if it be suitable for a home. The author has seen directly after a rain quite a "flock" of little mud-turtles half a mile from any body of water.

With the exceptions of the groups of turtles often seen sunning themselves on a log, and the bundling of snakes in hibernating, reptiles are not known to be very social or gregarious. However, sea-snakes go in shoals, as do some fishes. Reptiles may call to each other at the social season in some faint way— even the snakes having, perhaps, some little voice besides their hiss. It is fairly certain that mother-snakes signal in some manner to their young. The rattlesnake is said to use its rattle as a call to its mate, as well as a threatening implement or a warning. It is supposed by many students that this rattle is not in any way intended for the benefit of the foe, but for the snake's own good; that while animals are really warned of their danger in a way which they will heed, the snake also is thereby saved from a battle and the usual fatal consequences.

Though poisoned to the death, nearly anything that attacks a rattlesnake kills it before the poison takes effect. The warning is similar in purpose to that, of bright warning colors seen in the brilliant snakes, some stinging insects, and in Belt's little frog already noted. We have seen that many harmless snakes imitate a rattle by vibrating an ordinary tail against a leaf, whereby doubtless they hope to terrify a foe. One serpent has rough serrations (or teeth-like notches) on its side-scales; and by rubbing the folds of the body together back and forth a sort of mechanical hiss is made which is a threat—perhaps at times a call. Any snake makes a husky rattling noise as its coils rub past each other in its excitement.

Other reptiles have rude voices—there being, it is said, some evidence of vocal cords in the geckos. Crocodiles are well known to have voice, bellowing being a very loud form of challenge by the males; and the mother and young each have a distinct cry. The tortoise-forms hiss or breathe audibly when disturbed, and they perhaps have a slight voice.

In this connection it may be noted that many reptiles evidently play with each other, but it is usually in their courting-antics where this prevails. Turtles have mock-fights between the sexes, and the male alligator makes a silly exhibition of himself in the presence of his mate by either turning round and round on land or circling in the water, which he churns threateningly with his tail, to show her how he would treat a rival should one come near. Dr. Merriam reports two snakes seen in the act of rearing up and apparently playing with each other.

It is quite evident that lizards romp with each

other ; and one raised a pet may show a disposition to play with its master, like a pet squirrel. Play in all animals is largely a mock fight or chase, though this is not always the case.

While we have already noticed weapons, and thus incidentally fighting methods, a few more words here may be proper. Chameleons in their fights root each other about and bite, though their teeth are too small to inflict serious injury. The battle is long kept up. Sometimes the tail of the vanquished is snatched off and eaten by the victor. While this may be the case among some lizards, others take a grip and hold on with bulldog fierceness, as do most turtles. The fights of the tortoise-forms are ridiculously clumsy and ineffective. One may seize another's foot and hold on for a long time—the boys say "till it thunders"—but much of the battle consists in biting at each other's shells. Alligators dodge in front of each other head to head, each trying to get past the side of the other, so the tail may be used. When one has hit the other a few thwacks, the beaten one retires. To make this terrible stroke, the back-bone has the lateral spines long, as noted, to which great muscles are attached.

As mentioned, snakes of the same species do not seem to fight together, but one has been often known (where the species differ) to swallow the other. This has occurred where the victor was only two inches longer and the head of the victim had probably to be digested away before the rest could be taken in. In some cases this has happened when two snakes seized at once the same prey. Both held on and the larger

10

swallowed prey and all.   But in such cases as those where the blacksnake and other nimble snakes attack the rattlesnake, the victim is seized behind the head and often enclosed in the folds of its captor.   There is also an effort always to drag the victim in some definite direction, progress and resistance being affected by the tails of each curling about anything which can be reached.   It is a very exciting contest of skill, strength, and peculiar tactics.

ENEMIES

Perhaps there is no better connection than this to speak of the many enemies which the Reptiles have, besides man.   Tortoise-forms are so well protected that natural foes do not so readily get at them.   It is said by Pliny, however, that a Grecian poet was killed by a tortoise which an eagle let drop from a point high in the air—thus hoping perhaps to crush its shell.   Since it is well known that some crows— especially the fish-crows of our Northwest coast—rise thus with shell-fish and drop them to crush them on the rocks, this story does not seem so improbable. Vultures and other birds of prey are sometimes able to kill and devour a poorly protected turtle far from water.

Alligators when grown have few enemies except man, but their eggs and their young are eaten not only by nearly every creature in search of prey but by the old male alligators.   Lizards are eaten by birds of prey, wading birds, and others.   Snakes also eat them when they can catch them, and the lizards sometimes devour each other. In many cases their foes, like those of serpents, are legion. Cats, dogs, and many small rodents—especially the celebrated mongoos of India—all wading birds and birds of prey, and that remarkable half-and-half bird between these two groups, the secretary-bird of Africa—are enemies of snakes. Our little kites hunt for them

FIG. 58.—Gila monster of Arizona (*Heloderma suspectum*). Another species (*H. horridum*) inhabits Mexico.

and may be seen to eat them during flight. Man and hogs, of course, are the great enemies in civilized regions.

While all poisonous snakes should be killed, it is better in many cases where field-mice, gophers, ground-squirrels, etc., prevail too plentifully, to allow the non-poisonous snakes to live. There is only one poisonous lizard, the so-called Gila monster (see Fig. 58). With these exceptions, the small reptiles are not only innocent but useful, taking all things into consideration. Leeches live on turtles, often sucking blood even through the shell.

Age and Size

Reptiles, as a rule, are long-lived—especially the larger kind. These last appear to grow for a long time, but the small ones have a limit soon reached. Doubtless the others have also, late in life. Turtles have been found as long as six feet, and perhaps two hundred years old, even greater age being recorded; and they are said to be from eleven to thirteen years in maturing. The ordinary box-terrapin has been known to live fifty years. Alligators and the large snakes doubtless reach great age. Even the common tortoises live through two or three generations of man, since they have been kept that long as pets in family after family.

CHAPTER XIV

DIGESTIVE TRACT, RESPIRATION, CIRCULATION, LUNGS, HEART, COLD-BLOOD, LYMPH-HEARTS, SKELETON, SKULL, MUSCLES, NERVES, BRAIN, WISDOM, SKIN, AND SCALES IN REPTILES

The Digestive Tract

IN connection with long life there is nothing like good digestion. In the Reptiles the salivary glands first make their appearance—another thing which came in with them. Digestion therefore in them begins in the mouth, and that of the serpents is not excelled in all Nature so far as quickly dissolving large masses of flesh is concerned. Other creatures can excel them in digesting peculiar substances. Out of these salivary glands the poison-glands are made, which also, as noted, aid in digestion.

The digestive tract in all reptiles is rather simple—especially in serpents where it is very slightly twisted. In tortoises, however, and some lizards—and especially the vegetable-feeders—it is more complex. The liver is, of course, always present.

Respiration and Circulation

It is by the blood, after receiving oxygen from the lungs, that the elements of the food are carried into the system, thereby imparting energy and strength to the various parts of the body. It is no use to eat unless you can breathe, and no use to breathe unless the blood can circulate. Hence the grouping of these topics. Reptiles, as noted, breathe by lungs mostly, never by gills. To some extent the aquatic turtles breathe by the skin, though there is no such arrangement for this in these as there is in the amphibians. There are, however, near the skin some places where arteries are massed and twisted or meshed together, evidently to store aerated blood for use under water, if not for getting it further oxidized through the pores. This is the case, perhaps, in water-haunting tortoise-forms only.

LUNGS

The lungs in the average reptile are better than those of the average amphibian, but in many cases they still show a tendency to be mere sacs or pockets at the rear ends. With such imperfect lungs and a heart three-chambered only, we should still expect cold blood, as is the case in all living reptiles. In the old, very active *Dinosaurs* (of which more later) the bones were hollow and connected with the lungs; and there were, perhaps, other air sacs among the muscles, to help a small poor lung out in unusual work.

There can be no doubt that bones in birds and other creatures are made hollow to make them light and yet strong; but the air does not enter them to make them buoyant—notwithstanding that you will see such a statement made over and over even by fair authorities. One has recently said that a certain *Dinosaur inflates its bones* to make itself float better. The bones would be lighter if no air were in the hollow space. A balloon would soar better if no gas were in it, provided it were stiff enough not to collapse from pressure of the outside air. Nothing is lighter than *nothing*. The only advantage in buoyancy which the bones would have in getting air from the lungs is that it is warmer, generally, than that which might be infused from the outside, but this slight expansion from heat is not appreciable.

Lungs, when present, are paired in all creatures, except in the higher fishes, but in the long, limbless reptiles, whether snakes or lizards, the left is often a mere rudiment, and nearly always, except in one family, is noticeably smaller than the right. The *Amphisbœnidœ* are the only vertebrates in which the right lung is a vestige and the left one larger.

This remaining lung is long and slim to suit its space, and in serpents it runs far back. In the sea-snakes it lies along the entire body-cavity, is large and saclike at the rear, and stores such a quantity of air that the creatures can remain long under the water. In some reptiles the lungs are smooth sacs inside—not pouched or cellular at all anywhere.

The rule is that reptiles breathe by means of compressing and expanding the ribs only, but in the tortoise-forms these are fast to the shell, so that they breathe in about three different ways, hinting of frogs, birds, and mammals. They are the first creatures in the scale to have a muscle lie under the lungs and heart (called the diaphragm), by which the lungs can be slightly compressed. In some of them the lower shell and abdominal parts move generally to compress the lungs; and besides this, they get air into the lungs by swallowing it after the manner of the frogs. In those where the lower shell is fixed to the upper, the bones inside of it which form the shoulder are said to move in breathing. Their lungs are short and have many air-cells. A lizard breathes in the ordinary way, and so very rapidly that the expression, "Panting like a lizard," is proverbial. Snakes and lizards have no hint of a diaphragm.

Some lizards have expanded places in their windpipes in which they store air; and our common "spread-head" snake (*Heterodon*) has great lung-sacs that extend forward toward the head, to aid it, perhaps, in blowing or hissing. In the crocodilians there is such a large extra sac that there appear to be really three lung-sacs. These are very cellular, and all communicate with each other. They hold great quantities of air by which the creature can remain immersed for a long time.

CIRCULATION AND THE HEART

While Reptiles are always spoken of as having three-chambered hearts, some of the tortoise-forms have the partition between the two receiving chambers (*auricles*) not complete, and hence, since there is only one ventricle (pumping-out chamber), these hearts are really only two-chambered. In these, however, the old bulb below the heart in the vein, which is found in the fishes and amphibians, pulsates, and the one above the heart, situated in the artery, is likewise inclined to be active. The ventricle also tends to be divided so as to give a good circulation.

While the lizards, higher up, may have better hearts in some respects, they may have often poorer lungs, and hence the matter is about balanced. Some lizards show quite a tendency to have a partition in the ventricle, and the crocodilians succeed in doing so, and hence their hearts are four-chambered.

We recall in the low amphibians and the fishes that there are three or four branches running up on each side from the great tube which starts forward from the heart where the bulb throbs—one to each gill-arch. The frogs had three. Normally the Reptiles retain only one of these on *each* side (a few keep more), which seems to be a very symmetrical (well-balanced) arrangement. To show the onward step, we might mention that the mammals and birds keep only one of these and that on *one* side, the mammals retaining the left and the birds the right. This is the case also in the monitor-lizards and the chameleon, an onward hint, evidently, in this cold-blooded class. In the limbless *Amphisbænidæ* also there is only one. But in the crocodilians, although the heart is perfect, there remains two of these forward tubes; and after leaving the heart, they cross each other, and there is at the crossing an opening from one to the other. The rule is, that in other reptiles the blood comes from the system (through an auricle) into the single pumping-chamber, from which three tubes run—the two already noted as going forward into the body and the one running to the lungs. This latter soon forks—one branch going to each lung. Usually—especially in tortoise-forms—there is such an arrangement of openings and valves that much of this blood from the system is forced into the lungs, and this returns in such a way as to be driven out into the system anew and not back into the lungs again.

The description of these arrangements here is too technical for our purpose. In the crocodilians, however, where there are two pumping-chambers, the right has two tubes leaving it—one for the system and one for the lungs. Into the lung-tube the already used blood comes from the body. Here it will be seen that the lungs get much of the blood to aerate, but not all of it, since some goes on into the remaining artery which runs mainly to the lower part of the body. When the blood returns from the lungs, red and hotter, it reaches the other pumping-chamber and can not mix here with the colder blood. From this, only one artery runs forward, and it goes largely to the forepart of the body, or sends off many branches that way, the rest of it running to the rear. Now if it were not for the hole, where the two tubes cross outside of the heart, the forepart of the body would have well-aerated blood all the time and the rear part some hot blood, but mostly cool blood—a condition just suited to the position when the body is half out and half in the water. We saw that the frog had something like this; and it is quite probable that by such similar arrangements—as fulness of tubes, times of pumping of each ventricle, position of openings, and position of membranes or valves in the region—the crocodile, when breathing the air, is really warm-blooded in the front part of the body and cold-blooded everywhere else, but, by means of this leak, he is wholly cold-blooded when he is under the water holding his breath. At least some students have so thought, and the purpose of this hole implies something like it. If true, it is one of the most wonderful cases of adaptation to variable conditions found in Nature.

To be wholly hot-blooded the crocodilians would have to do away with the artery running to the body from the right ventricle, and then, of course, close the hole in that running from the left. Then the condition would be similar to those in the birds and mammals, and the four-chambered heart would mean more than a mere prophecy. Had both arteries run from the left ventricle, however, and the hole between them remained, the animal would still have been hot-blooded, if such lungs could have made it so.

The author trusts he may be pardoned for really trying to show the different causes which may bring about hot and cold blood, since most popular works say little about it.

Lymph-Hearts

In reptiles there are lymph-hearts at the root of the tail as in amphibians and some birds; and in all these low creatures and some mammals, there are places—usually in the armpit—where the blood-veins pulsate independently as did those which first formed the heart in the lowest Vertebrates. These show readily in the wings of bats.

Skeleton Generally

Much has been said already of the bones of the reptiles as we have come along. This class is more

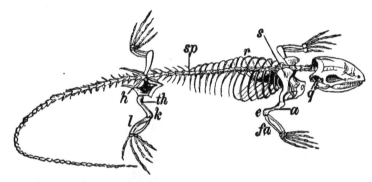

FIG. 59.—Skeleton of a lizard. *sp*, spinous processes, which in the tortoise are flattened into plates; *r*, ribs; *s*, shoulder-bone; *a*, upper arm; *e*, elbow; *fa*, forearm; *h*, hip-bone; *th*, thigh-bone; *k*, knee; *l*, bones of the leg; *q*, quadrate bone between upper and lower jaw.

bony than the amphibians, though some of the ancient forms had the old soft gristly string yet inside of the rings of bones which made up their spinal columns; and the gecko tribe (Fig. 60) has something similar yet. In geckos the vertebræ are cupped at both ends; in most others they are cupped back and round in front. In the old *Dinosaurs*, however, the vertebræ were often cupped in front, which is a higher form found frequently in the birds and largely among the mammals.

Reptiles and birds usually make the shoulder junction out of three bones, while mammals have only two. They omit the one that floats loosely back over the side, called the *Coracoid*. While nearly all reptiles have a collar-bone (or bones) the flying *Pterodactyl* forms had none.

In the thigh-junction, the Reptiles have peculiarities which the anatomist easily recognizes. In such

Fig. 60.—Wall gecko (*Platydactylus muralis*). *A*, foot of gecko, showing suction slats by means of which it adheres to perpendicular surfaces.

reptiles as walked on two legs these bones have quite a birdlike arrangement. The *Dinosaur* and bird have long projecting bones behind to support them in an upright posture.

The RIBS have been noted as many and long—

usually reaching the breast-bone when that is present. In many ancient forms the ribs appear short, and there is no evidence of a breast-bone. In crocodilians and *Tuatera* the ribs have, near the middle, projections or flat blunt spurs (uncinate processes) which reach back and lap over the next rib, thus strengthening the bony cage. The lower birds also have this.

In tortoise-forms the ribs are usually fused into the shell and are broad and flattened. In one case,

FIG. 61.—Tortoise-shell turtle (hawkbill) (*Eretmochelys imbricata*).

however, the sea-turtle (*Sphargis*), the ribs are free from the shell but much flattened. It has been claimed that the tortoise-forms have no true ribs, but that those apparent are merely the spiny projections of the back-bone; but they, at least, take the place of ribs, and in this sea-turtle noted their freedom hints that they really are such. So likewise tortoise-forms have no breast-bone, but their lower shell takes its place, though the bones of this shell have their origin in the skin. There are many peculiarities in the ribs of reptiles which we can not even mention.

Those of the serpents and flying lizards have been noted under MOTION. In crocodilians the breast-bone runs back to the thigh-junction, and peculiar ribs, loose at their upper ends, are attached to it; and in the *Tuatera* there are ribs in the abdomen not found connected with anything else. Perhaps the slightest hint of the lower shell of the tortoises may lie in this.

The SKULL of the reptile is more complete than in the amphibians—that is, the box is more perfect —better roofed-in. In all cases but one (which is not normal) it joins the first vertebræ of the back-bone by only *one* ball-and-socket joint—the ball being on the skull. The skull is much like that of birds and some very low mammals—as the opossums; but from these last, it is sufficiently distinct, if we could specify without being tedious and technical. The lower jaw's connections to the skull have been noticed.

MUSCLES

MUSCLES of reptiles are especially interesting to students because of what they hint of the origin of peculiar muscles in birds and mammals; but the subjcet is too broad to discuss here. They all tend to be redder than those of the amphibians—having a better blood supply and are thus more active; but they are not so dark as those of the more agile, more red-blooded birds and mammals.

We can see from the great surfaces on the bones of the old monsters that there were some tremendous muscles among reptiles in those days (Fig. 62). In such as flew there was a great breast-muscle, like that in birds, which worked the wings, and some had a high ridge or keel on the breast-bone. In all cases the ribs have more muscles in the reptiles than in the mammals.

THE NERVES AND BRAIN

Many reptiles as well as the amphibians, as already noted, have a great deal of their nervous system outside of the skull, and hence have small amount of *cerebellum*—that part of the brain which moves the muscles. They therefore die hard after the head is cut off. A turtle is one of these, and is said to be able to live eighteen days after the brain is removed. Its movements are then, of course, largely influenced by reflex action. Serpents also are tenacious of life, and squirm long after the brain is destroyed. One with its head off will sometimes turn and strike at any disturbing thing on the body. The author has a friend who nearly had the wits scared out of him by a headless rattlesnake striking his hand as he attempted to pull the rattles from the tail. In this connection it may be well to notice that in some old *Dinosaurs* there was a hollow expansion in the back-bone near the hips that was much larger than the skull-cavity and held many times more nerve-matter. It was a sort of second or lower brain. Here is Professor Marsh's restoration of one of these—the *Stegosaurus*—with shellplates set on edge. Look at the small skull. The brain was quite likely used only for the senses of

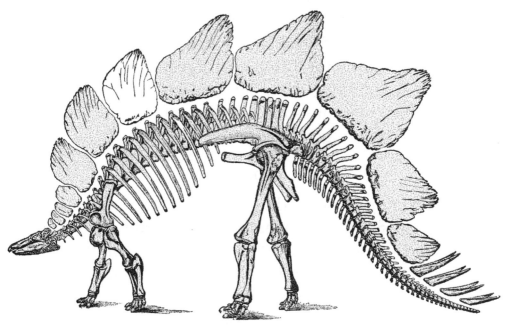

FIG. 62.—*Stegosaurus ungulatus* (after Marsh), × 1/80, Jurassic, Wyoming.

sight, smell, etc., and it is not unlikely that it would have been dangerous to come within reach of that terrible spiny tail months after the head was cut off (Fig. 62).

We can not note the parts of the brain except to say that a certain little gland found on the top of that of mammals is observed to be much larger in most reptiles, and it was once, doubtless, in some, an active third optic lobe corresponding to a useful third eye.

11

In *Tuatera* there are parts of this eye yet remaining —but they are not now useful.

Reptiles are more intelligent than amphibians, and have a certain form of cunning in obtaining prey and effecting escape from enemies; but so far as any great amount of intellect is concerned they do not compare with birds and mammals. The wisdom of serpents—so far-famed in the minds of Eastern peoples—is used in the Bible as a figure, not because it was such a striking fact, but because it was a familiar method of expressing that sort of cunning which is devoid of any good attribute.

SKIN

Covering all these things that we have been noticing is the SKIN. We mention it here only in connection with the renewal of the outer part of it. No creature changes its real skin any more than its other tissues. In the lizards the transparent thin outer membrane runs outside of the scales, and even outside of the eyes in the snakes. The former usually shed this in strips; but some limbless lizards and all snakes, except the sea-snakes, shed it in one piece, which is a fine image of the creature. In many snakes and some lizards a series of small hairs seems to arise and push the epidermis loose. In some horned toads (the spiny lizards out West) there arises beneath the outer skin small watery pimples (pustules) which push it loose till it breaks off in little pieces; so that this homely creature has to have an attack of chicken-pox every time it gets a new suit.

In snakes the eyeballs can not be moved beneath this outer skin. In rattlesnakes every time this skin is shed there is left a ridge or rim of it at the tail, which forms a rattle. The first remnant is called a " button." The skin may be shed more than once a summer so that the number of rattles may not tell the age in years, but they form a means of rough comparison. As already noted, the folds of the thick skin proper become horny and form the scales ; and thus the skin has built all the outer coverings for the reptiles, and many bones that are beneath it besides.

CHAPTER XV

SENSE-ORGANS

ALL the senses are closely related primarily with the skin and may be taken up next in order. Except that of TASTE, the senses of the Reptiles are quite keen. Even where such a tongue as that in the tortoise-forms prevails along with the habit of chewing the food, taste may be considerable. But in such reptiles as serpents, which gulp their food, there can be little enjoyment of it except that of fulness and a sense of good digestion.

Serpents and many lizards use the tongue as a feeling organ. Besides this, the reptiles are not largely endowed with organs of TOUCH. The tortoise-forms, however, know at once when their shells are touched never so slightly, and scales may be equally sensitive. One turtle—the *Matamata*—has a long fleshy snout, and above and below the neck there are a series of filaments, but it is likely that these latter are made to resemble seaweed, and are

138

therefore intended more for concealing or protecting than for feeling.

The HEARING of reptiles is quite acute. Though they have much better ears than any creature below them these are far from being so complete as those of the mammals or even those of the birds. The ears are connected with the mouth by tubes—in the crocodilians by three openings. This forms the internal ear. Besides this there is a middle ear of more or less incompleteness. Except in crocodiles, where there is an external fold or flap of skin, there is no outside ear. Here there are some valves over an opening to keep the water out, but in many reptiles the drum-membrane lies on the outside of the head, as in the frogs, except that it may be protected by special scales. In others, as the serpents and *Tuatera*, there is no drum-cavity; and in the latter only, it is said, are there any signs of that wonderfully spiral arrangement called the *cochlea*, which allows us to appreciate the difference between high and low tones in pitch.

All reptiles *see*, there being none known that are perfectly blind, but some burrowing forms are nearly so. SIGHT in this class, however, has no such keenness as in the birds. Reptiles usually have two eyelids, and some have that third kind called a nictitating membrane. The skinks have a transparent window in the lower lid, so that while burrowing slightly in deserts they may see out without getting sand in the eye. Tear-glands here prevail for the first time in vertebrates. As noted, snakes and some lizards have the eyelids covered by the outside skin so that they can not be moved.

In the chameleons the eye-opening may be round and is drawn up like an old-fashioned purse—as if a draw-string were around it. One eye can be turned forward while the other is rolled backward, so that the creature can really look two ways at once. This peculiarity is not known anywhere else, though birds and many mammals see two ways at once because one eye is on each side of the head. But both eyes move forward or backward together.

We have noted that *Tuatera* has beneath its skull the remnants of a third eye, which was once central in the top of the head. Some other lizards show hints of the same thing; and in the growth of the young of all, there is in the roof of the skull a place that is slow to close. This has been supposed by some to be the vestige of the former opening of this third eye.

The fact that reptiles are so odorous argues that they SMELL, if there were no other means of inferring it. All breathe air through the nostrils which, of course, unlike those of fishes, enter the mouth. In the crocodilians these enter far back, and there is a soft flap of the palate which drops down in front of them so that the creature can keep its mouth open under water and not get strangled; and at the extreme end of the snout, where the nose opens outward, are valves which can be closed to keep the water out.

In the water-snakes, also, the nostrils open near the tip—really on the upper side of the snout—so that they breathe easily when near the surface without thrusting the head far up.   In one of the croco-

FIG. 63.—Gavial (*Gavialis gangetica*).

dilians—the Gavial (Fig. 63)—the male has a great warty bunch on the snout through which the nostrils open.

EGGS, HATCHING, AND CARE OF YOUNG

We have spoken of the eggs of reptiles as being large and free from each other—like those of birds. Those of the crocodilians and all tortoise-forms, except the paddle-feet kinds, have limy hard shells, but in all others a tough membrane only covers them. They consist of a yolk and "white" as in hen eggs, and in no case are they fertilized (or made so they will hatch) after they have been laid, as is the case with most fishes and some amphibians. In shape they are all longer than thick, with both ends equally rounded, but a few turtles' eggs are almost as globular as a boy's marble. In no case are they pointed more at one end than the other, and, so far as known, they have never possessed such special colors as are found in those of birds.

In nesting, tortoise-forms bury their eggs under sand—scraping a place and covering them. The rule is that the mother takes no further care of her young. The sun's heat alone hatches them.

In the crocodilians it is said that there is some watching of the buried eggs; and the mother digs her young out when hatched, leads them to the water, fights the male away, who would eat them, it is said, and in some cases she ejects food from her own stomach to feed them. Some of this order dig deep holes and put trash or litter of vegetable substances in with their eggs and then cover all over. It is claimed by some students that this procedure is for the purpose of getting additional heat by the hot-bed effect of the rotting litter; by others, that it is intended merely to cushion or protect the eggs—perhaps a sort of nest-building instinct. Certain it is that some low birds bury their eggs in almost exactly the same manner, and that others of the same order heap up a great mass of brush, grass, etc., and lay their eggs in this, where they are hatched either by the heat of decay or by that of the sun.

One large poisonous snake of Asia is said to make a nest of a heap of rubbish, and to defend it—even pursuing the disturbing enemy. But the usual snake either burrows under loose earth or into the rotted roots of old stumps and other crevices, and there deposits her eggs. Sometimes they are not at all or scarcely hidden.

In all reptiles there are many eggs at a litter. In the constricting families of snakes—especially the pythons—the mother may coil herself about her eggs, incubating them as a hen sits on hers to hatch them. At such times it is found that the temperature of the cold-blooded mother is raised very noticeably—a thing which is well known to occur in brooding birds. In some snakes the eggs are hatched within the body and the young born perfect.

That certain snakes care for their young there can be little doubt; whether all do or not can not be asserted. The author is convinced from reliable witnesses that mother snakes may take their young into their mouths, gullets, and stomachs even, in time of danger, and that they have some sort of call which warns of approaching peril. It has been thought by skeptics that since some snakes are viviparous, the unscientific persons find the young before they are born and are thus deceived; but such trustworthy accounts state that the young are seen to enter, are found in the *stomach* proper, and have been pushed out of the mouth after death from snakes well known to be egg-laying.

Lizards lay eggs in crannies and under the soil. Some of these also bring forth their young alive. There is little evidence that they care for their young.

All reptiles are hatched or born perfect and active (*precocial*). There is no imperfect form, as in amphibians and some fishes—not even as much as there is in many birds and mammals. Even if their eggs are broken open a little prematurely (before being just ready) the young will often escape open-eyed and active at once, and survive. Little snakes and others have a tiny hard shell or "pip" on their noses—as in chickens—with which they cut the shells at the proper time.

With this large egg of the reptiles and the perfection of the young within it (not out of it) there came in two membranes which enclose the young, called the *amnion* and the *allantois*, but these are connected with the nourishing and the breathing of the little creature in a manner too technical for our study. They are not found in any degree of perfection below the reptiles and the birds.

In HATCHING, the very beginning of the little reptile is much the same as that of the amphibian and fish, but each order soon hints its peculiar shape. The turtle quickly shows its short body and shell with the ribs free at first; the lizard-forms show the limbs, and the snake its peculiar lithe form. In no stage does the snake show any more sign of limbs than it does in the adult—a fact which implies that it has been for a long time only a snake. The study of these developments is the science of embryology—a whole realm of thought all by itself, which has made some wonderful suggestions at kinships and ancestors in all classes.

Reptiles in Geology

Except the *Tuatera* all our modern reptiles belong to families that are comparatively recent. The
crocodilians, however, had ancient representatives.
If this order were not one of the families of the old
order of *Dinosaurs*, it was close to them. They
occur just above the coal—after the age in which the
Reptiles all came in. It is probable that they are
now (and were then) a slightly degraded form, and
got the hole between the two great heart arteries by
going back to the water after having been once
more terrestrial.

The tortoise-forms are not found farther back
than the rocks which are just below man—the so-
called Tertiary times. They are evidently just lizards
with shells on the back. As noted, many lizards show
bones on the back under the skin, and other lizard-
forms have in the abdomen bones which are not
part of the breast-bone; some have beaks quite like
those of the turtles, but the teeth are still present.
One turtle, however, shows a tooth or two in its very
young. Finally, a small fossil burrowing lizard has
been found by Professor Cope which had a shell on top
only. A giant fossil turtle was found out West thirteen feet long and it had flippers which measured from
tip to tip about fifteen feet. Like the sea-turtle its
ribs were flattened but were not grown to the shell.

Snakes are found in the lowest Tertiary—perhaps
a little older than the tortoise-forms. They are but
lizards with the legs lost and the chin loose-jointed.

Later, other lizards lost their legs also, but they did not go in the snakeward direction in other respects.

Lizards themselves are the basic order, and they had many different representatives in the long-ago. Along with the *Tuatera* and its order, and the crocodiles of the old rocks, the lizards had some very modern-shaped forms in ancient times; but the most interesting kinds were such as were peculiar to those ages—the giants of those days.

It is remarkable that in this group of great monsters Nature ran nearly the entire scale of present vertebrate shapes as though she were trying her hand roughly, or as if the surroundings then were forcing her soft and pliable materials into the type-forms of the future. Here were fish-forms, serpent-forms, the true or typical reptile-forms, the beast- or mammal-forms and the bird-forms. This last took on two manifestations—one in the fore limbs, the other in the hind limbs. Roundly we may say these monsters were sea-haunters, land-haunters, and air-haunters.

If we draw circles to represent kinships we shall see that the reptiles really touch all the other groups and may cut into them a little way (Fig. 64).

We have already spoken of the *Theromorphs*, whose teeth at least were so much like those of mammals. It does not follow that they were the ancestors of our present carnivorous beasts, but the same conditions produced these then which later made the mammals. They were, however, among the earliest reptiles, and had in them the possibilities of all things. They had in them elements which are found in all other groups modern and extinct.

The next three orders of these monsters were aquatic purely—degenerates probably from a terres-

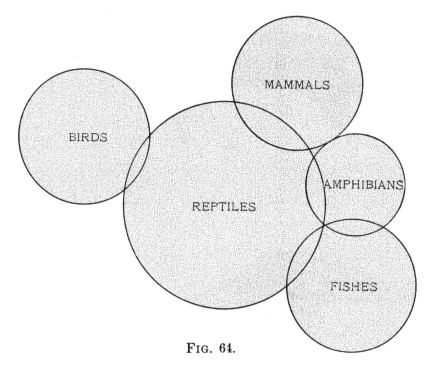

FIG. 64.

trial form, and were something like the crocodilians, though they are often thought to come up from the fishes. They all had paddles for limbs, and had tails adapted to swimming. Professor Cope makes three very natural orders out of three forms of them.

The first, *Sauropterygia*, had whalelike bodies with necks longer than the tails. The *Plesiosaurus* (Fig. 65) is a type of these.

The next group of paddle-limbed kinds is called *Ichthyopterygia*, because they had a sort of fishlike body (Greek *ichthyos*—a fish) and a long pointed tail

It is these also which had the most finlike paddles. The neck was very short (Fig. 66). Excepting the paddles they resemble alligators in general shape quite strikingly. Some later forms were

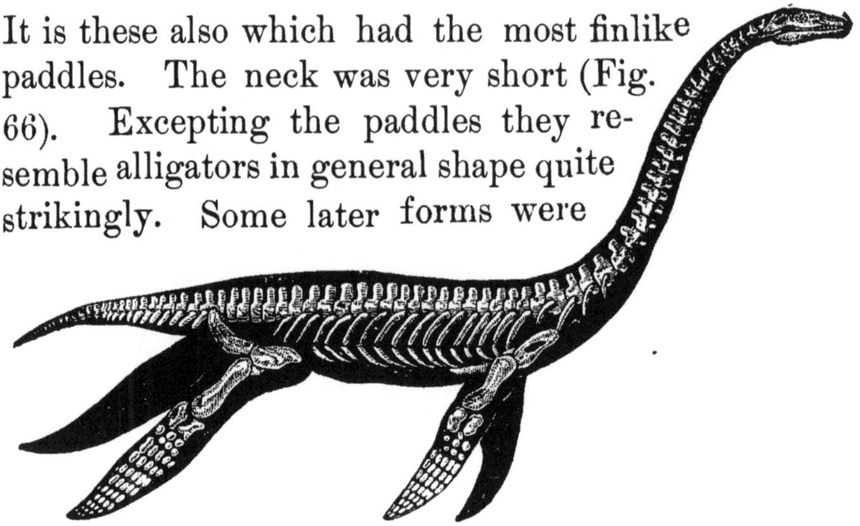

FIG. 65.—*Plesiosaurus dolichodirus*, restored, × $\frac{1}{80}$.

toothless. It is in these that the six digits in the feet and hands are found (Fig. 67).

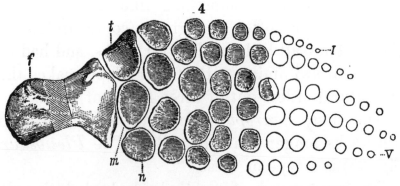

FIG. 66.—*Ichthyosaurus communis*, × $\frac{1}{100}$.

FIG. 67.—Left hind-paddle of *Baptanodon discus* (after Marsh), seen from below. One-eighth natural size. *f*, femur; *t*, tibia; *i*, fibula; *I*, first digit; *V*, fifth digit.

The third one of these paddle-limbed orders had bodies that were snakelike with extremely long tails. The type‑form is the *Mosasaurus*, found fossil in the Old World; but America also is rich in them. Professor Cope has called them *Pythonomorpha*. One of them—the *Edestosaurus*—is here shown (Fig. 68). It will be noted that their limbs, while paddlelike, had the digits more like those of land-haunters. It can be seen by the lengthened spines on the tail vertebræ that the tail had a crest and was a swimming organ. They are much like serpents in some of their parts. It was in these that the jaws were doubly hinged for swallowing. Many had more ribs than the one here figured, and there was no breast-bone meeting of ribs below.

The next division is the great land-haunting group known collectively as *Dinosaurs*. It is supposed (and probable) that the crocodilians lay between the *Ichthyosaurus* and these. Here Nature seemed to have tried all things and hoped all things—holding fast to nothing. We can note only a few typical forms.

While some were four-footed in

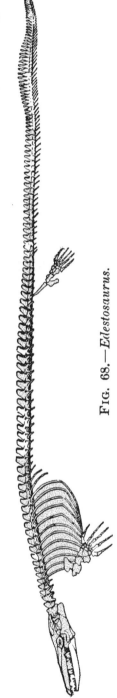

FIG. 68.—*Edestosaurus.*

their motion, perhaps all were capable of rearing up, many of sitting up permanently, and a large proportion could move about on the hind feet with or without touching the tail to the ground. In all, however, the tail was an important member—the third leg often of a three-cornered support. The *Stegosaurus*, already described (see page 135, Fig. 62) as so well protected and armed in the tail, and the *Triceratops*—which was three-horned (Fig. 69)—are examples of

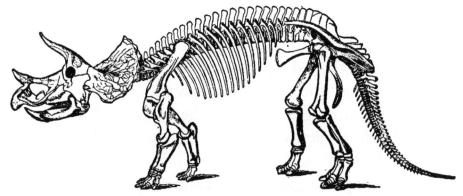

FIG. 69.—*Triceratops prorsus* (after Marsh), × ᵃ₃₀, Cretaceous, Wyoming.

the four-footed walkers of this group. This last is the one which had a collar of spines also. It had a beak which was tortoiselike in some respects. The next one, *Brontosaurus*, shows by its small fore limbs (Fig. 70) that it arose on its hind legs; but a glance at its neck bones and the set of its skull shows that its head was held horizontal by the spines of the backbone. Observe that the elongation of rear limbs,. the slim neck vertebræ, and the long head, set at an angle with the neck, show that *Laosaurus* (Fig. 71) walked erect. All of these noted were herbivorous—browsers on tall trees. Some were able to feed thirty and forty feet up.

One was in the habit of wading on the bottom of

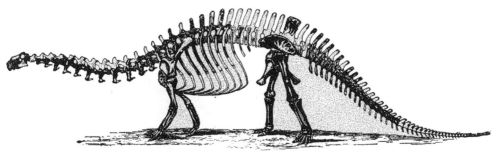

FIG. 70.—*Brontosaurus excelsis,* × $\frac{1}{120}$ (restored by Marsh).

deep pools and projecting its long neck above the water and in all probability was quite helpless on

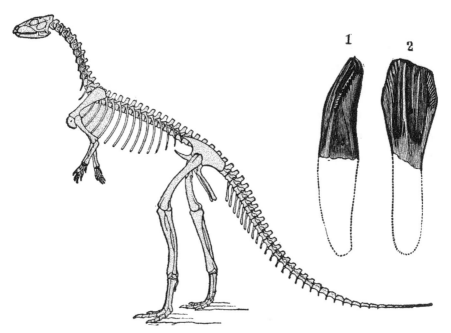

FIG. 71.—Restoration of Laosaurus by Marsh, $\frac{1}{20}$. 1, tooth of *Lao-saurus altus* (after Marsh), front view; 2, the same, side view. Both twice natural size.

land, on account of its weight. Another, with ordinary feet, had a bill like that of a spoonbill or "shoveler" duck, and fed only on more tender aquatic plants, as is shown by its teeth (Fig. 72). Upon this the terrible *Lælaps*—the talon-toed flesh-eater—is supposed to have fed.

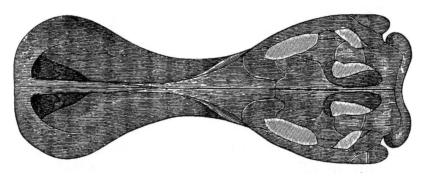

FIG. 72.—*Diclonius mirabilis* × ¼ (after Cope).

If we glance at Fig. 73, on page 153, we may note how birdlike these reptiles were in shape, yet the resemblance was stronger still in structure. It is not necessary, however, to believe that birds came from these monsters; but there were some strong birdward tendencies at work then. On page 154 are the cuts of some tracks made by these old three-toed bipedal walkers (Fig. 74).

One of the most peculiar of these giants has been called the fin-backed lizard. It was aquatic, swimming by a long flattened tail. The spines on the upper side of the back-bone grew up into a great high arch like a fin; and these, quite likely, had membranes over them or between them. On one of these, found by Professor Cope, there were cross-arms as on the mast of a ship. We can see no use for these, unless there was a cross membrane also; and the creature may have thereby actually sailed before the breeze. Those that did not have the cross-arms

FIG. 73.—*Hesperornis regalis*, × 1/10 (restored by Marsh).

may have tacked by means of the single spread of the broad back-fin. This latter would be useful, however, as the dorsal fin of a fish is, in keeping the body steady while the tail lashed (Fig. 75).

The oldest real reptile known was found by Professor Cope in Ohio, away below the level where reptiles are usually found. It is quite like a salamander in shape; and doubtless from something like

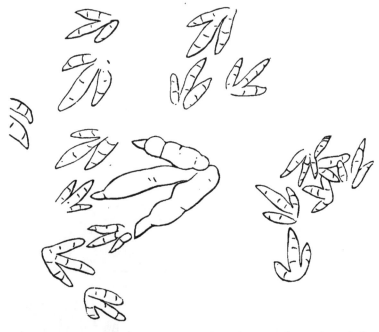

FIG. 74.—Portion of a slab with tracks of several species of Brontozoum (after Hitchcock).

this all those monsters and our living forms have been developed. Of course there is no space here to go into all the curious forms of reptiles found in the rocks, but the general types have been indicated.

The last order is that of the flying lizards already mentioned and illustrated (see page 114, Fig. 57). On page 156 is a cut also of the skeleton (Fig. 76). As noted, some others had very long tails. Some had teeth in their beaks; some had none. While they were so birdlike in appearance they show by their structure that they are not the ancestors of our birds,

FIG. 75.—Finback lizard floating on the surface of the water, borne along by the wind.

but were just a thrust of soft material into the birdward direction. In some way they were not fitted to survive, and with the rest of the freaky experiments of Nature they died out. We can scarcely guess at the cause of their disappearance. Perhaps their immense size finally made them too cumbersome to procure food easily, especially if the climate changed in their region—and those were changing days.

Their extermination by the more active mammals which followed them seems hardly probable, since so many of the reptiles were aquatic and so few, if any, mammals had then taken to the water. Many may have fallen before great numbers of small foes, just as whales and sharks perish now; or some special parasite or disease may have attacked them.

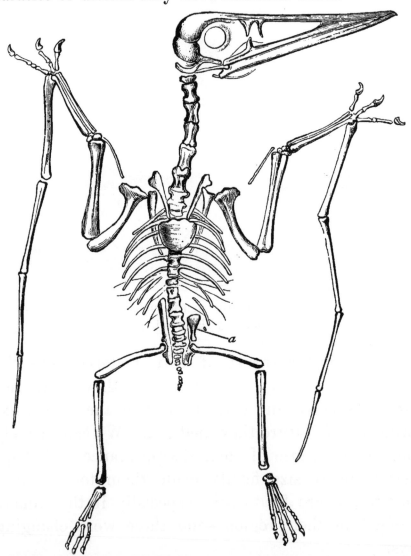

FIG. 76.—The nearly entire skeleton of *Pterodactylus spectabilis* (Von Meyer), as shown by the two halves of a split block of lithographic slate. *a*, the left prepubic bone; on the right side this bone is not shown, and the ilium is exposed.

Here is a brief review in the form of a key which will help to recall the *orders* only, of these old terrors of the past.

A. Jaws either toothless or beaklike; or else with only a pair of tusks or canine teeth evident.   *Theromorpha.*

AA. Jaws not so.   More teeth than two tusks always present.

  B. Limbs paddle-shaped.

    C. Body snakelike.   (Tail longer than neck.)   .

                              Pythonomorpha.

    CC. Body not snakelike.

      D. Neck very short—much shorter than the tail; body fishlike.   *Ichthyopterygia.*

      DD. Neck longer than tail; body whalelike.

                              Sauropterygia.

  BB. Limbs not paddle-shaped.

      E. Hind legs longest and largest—often fitted for walking erect; outer finger not elongated.

                              Dinosauria.

      EE. Hind legs *not* longest; outer finger elongated —fitted for flight.   *Pterosauria.*

CHAPTER XVI

OUTLINE OF MODERN REPTILES AND SOME HELPFUL KEYS

To help recall the peculiarities of the various orders of living reptiles the following outline in the form of a key is submitted, presenting apparent as well as structural features:

NOTE.—If one peculiarity is noted at a certain letter the opposite will occur where the letter is found doubled.

A. Form tortoiselike; a distinct shell. Shell either leathery or horny; mouth beaklike—no teeth; back-bone and ribs fast to shell (one exception). *Tortoise-forms.*

AA. Form not tortoiselike; either lizardlike or snakelike; jaws with evident teeth; back-bone always movable everywhere.

B. Tail with a double crest near the body; skin covered with plates placed edge to edge; heart four-chambered; toes four behind, five before. *Crocodilians.*

BB. Tail, toes, skin, etc., not as above; heart never four-chambered.

C. Teeth in upper jaw appearing to be in two rows—the lower teeth shutting between. *Tuatera.*

CC. Teeth in upper jaw not appearing to be in two rows, though teeth may be present on palate.

D. Lower jaw solid at chin—not jointed or loose. Limbs present or absent, but there is always a shoulder-girdle. *Lizards* proper.

DD. Lower jaws held together at chin by a ligament merely. No true limbs with toes. No shoulder-girdle; eyelids never apparent. Tongue always forked, slim, and sheathed. *Serpents.*

158

Legless lizards may be usually known from the snakes by the tongue, which is not in their case forked and threadlike, but flat and barely notched. Where lizards have a threadlike forked tongue, limbs are present. Eyelids and ear-drum membranes also are usually apparent here and not in snakes.

TORTOISE-FORMS

The following little outline will enable any one to know the families of tortoise-forms. It is based on shells and toes and is not just the natural order of kinships, but is very apparent as a means of describing :

A. Shell soft, leathery outside.
    B. Back ridged ; limbs as flippers.   *Leather Sea Turtles.*
    BB. Back not ridged ; limbs as legs.   *Soft-shelled Turtles.*
AA. Shell hard, with horny scales outside.
    C. Limbs as flippers or paddles.   *Loggerhead Turtles.*
    CC. Limbs as legs.
        D. Tail long, crested, or saw-toothed on top ; bottom shell not a complete disk.   *Snapping Turtles.*
        DD. Tail short, not crested ; bottom shell broad.
            E. Toes free or with webs only between.
            F. Twelve divisions in lower shell.
                Pond Turtles.
            FF. Not twelve divisions in lower shell.
                Box-Tortoises.
            EE. Toes not free ; all inside one mass of flesh ; claws only showing.
                Land-Tortoises.

The hinging of the shell can not be used in separating families. There are " box-turtles " in both the pond-turtle family and the box-turtle (*Kinosternia*) family proper, but the two groups have very different structure. There is quite a tendency now to classify the order by the peculiar jointing of the bones of the neck—usually shown in the manner by which the creature tucks away its head in the shell, whether by drawing it directly in or by bending it sidewise. Perhaps this and other anatomical features—especially those connected with the head—may change our groups any day according to the emphasis put upon any one set of peculiarities.

The Crocodilians

These may be roughly divided as follows:

A. Tusks of lower jaw bite into *grooves* of the upper.
      (Crocodiles generally.)
   B. Edges of jaw nearly straight; jaws long, thin, and narrow.
                              Gavial.
   BB. Edges of jaw scalloped or wavy.
                        Egyptian or *True Crocodiles.*
AA. Tusks of lower jaw bite into a *pit* in the upper.
      (Alligators generally.)
   C. Plates of back not hinged; simply connected by the skin.
      D. Bony plates on underside of body; front toes free.
                              Asiatic Alligator.
      DD. Bony plates wanting below; front toes webbed.
                        North American Alligator.
   CC. Plates of back hinged to each other.
         Caiman or *South American Alligator.*

Serpents

Dr. Günther has noted that in the popular mind snakes are divided by habit into—

sonous snakes—i. e., snakes with poison-fangs—which we have.

The little coral-snake of the Southern States is a poisonous snake, but its fangs are short and far back, and serious results rarely follow its bite.

Since the saliva of any creature may prove to be poisonous, especially if it be worried into a frenzy or be diseased, it is simply folly to allow it to bite you. Such assertions as those that "blacksnakes," "milksnakes," "garter snakes," "spread-heads," etc., have been known to kill, can only be believed upon the same grounds that rat-bites are sometimes fatal.

LIZARDS

We can not classify the lizards by habit either. Thus, the legless blindworm of Europe is now known to be rather near to the "Gila monster" (*Heloderma*) both in structure of body and the mutual possession of grooved teeth. The following key, though not practical without dissection, is a fair presentation of many peculiarities of the families the world over:

A. Vertebræ cupped at both ends.
    B. Collar-bone expanded into a loop at one end.
                                 (Gecko-forms.)
    C. No eyelids.   (1) *True Geckos.*
    CC. Eyelids present.   Another form of geckos.
                          (2) *Eublepharidæ.*
   BB. Collar-bone not expanded, etc. (More geckos.)
                       (3) *Uroplatedæ.*
AA. Vertebræ not cupped at both ends.
    D. Feet as merely projecting scales.
               (4) *Scale-footed Lizards.*
   DD. Feet ordinary or none. May be either two or four.

E. *Right* lung largest or active.　　(12) *Amphisbænidæ.*

EE. *Left* lung largest or more active.

　　F. Skin distinctly warty—not scaly; teeth grooved.

　　　　　　(9) *Helodermatidæ.　Gila Monster.*

　　FF. Skin not warty; either scaly or smooth.

　　　　G. Fold of skin *obvious* along side of body.

　　　　　　　　(7) *Girdled Lizards.*

　　　　GG. Fold of skin not obvious or present.

　　　　　　H. *Tongue distinctly forked* or with two points at tip.

　　　　　　　　I. Tongue scaly, rough, or wrinkled above.

　　　　　　　　　　Teiidæ or　　(11) *Greaved Lizards.*

　　　　　　　　II. Tongue not scaly or wrinkled; sometimes merely brushlike.

　　　　　　　　　　K. Tongue sheathed at base. *Varanidæ* or

　　　　　　　　　　　　(10) *Monitors.*

　　　　　　　　　　KK. Tongue exposed everywhere.

　　　　　　　　　　　　(13) *Lacertidæ.*

　　　　　　HH. *Tongue not distinctly forked,* or double-pointed; sometimes merely notched.

　　　　　　　　L. Scales of body *underlaid,* and roof of skull behind overlaid, with flat bones; scales not arranged in rows.

　　　　　　　　M. Tongue sheathed at the tip.

　　　　　　　　　　(8) *Blindworm Tribe, Anguidæ.*

　　　　　　　　MM. Tongue not sheathed at tip.

　　　　　　　　　　(14) *Skink Tribe.*

　　　　　　　　LL. Scales and skull devoid of the bones noted; scales in rows, usually oblique.

　　　　　　　　　　N. Teeth on upper edge of jaws.

　　　　　　　　　　　　(15) *Agamidæ.*

　　　　　　　　　　NN. Teeth on sides of jaws.

　　　　　　　　　　　　(16) *Iguanidæ.*

The numbers before the family name refer to the order of Mr. Lydekker's families.

The *Agamidæ* are a large Old World family in

which are the beautiful tree-lizards, flying dragons, Moloch Spring lizard, etc.

The *Iguanidæ* have interesting Old World members, but in South America there are the sea-lizards around the Galapagos Islands which are large vegetable feeders and peculiarly interesting. Our so-called chameleon (*Anolis*), the frilled lizard, and our horned toad are all in this family.

The family now known as *Anguidæ*—holding the blindworms—was formerly put with the skinks. It contains our so-called jointed-snakes.

The family known as *Varanidæ* contains the monitors—the largest of true living lizards. They are Asiatic, and some have extremely long tails which with the body measure nearly eight feet. They use the tail as a lash and strike a painful blow with it, and they fight viciously otherwise. Some are aquatic.

The *Teiidæ* are called greaved lizards because the head is covered with scales arranged after the manner of the old greaved armor. There are some of these in the United States.

The *Amphisbænidæ* have been noted for their burrowing habits, snakelike, limbless bodies, and their ability to go backward as well as forward.

The *Lacertidæ* hold the usual Old World lizards, whose tails come off and regrows so easily.

The *Scincidæ*, or skinks proper, are widely spread over all the continents. We have some. Our blue-tailed lizard and ground lizard belong here. Some in the Old World have only two legs and some none.

Of course chameleons proper form a family of lizards, but they differ so much that they have been put into a separate suborder by some students. There are many different kinds ; but all may be known from other lizards at a glance by their circular eyes and the bunching of the five toes into two fingers or toes opposing three others—there being no such eyes or feet anywhere in Nature.

The *families* of the lizards found in the Northeastern United States may be quickly determined by the following little key—if the specimen is in hand :

A. Limbs practically absent (in America).   Tongue not snakelike.
                   Anguidæ, Glass-Snake family.
AA. Limbs present.
    B. Tongue thick, not notched or scaly.   *Iguanidæ.*
    BB. Tongue thin, scaly, slightly notched.
                         Skinks or *Scincidæ.*
    BBB. Tongue broad, wrinkled or scaly, but ending in two
        sharp points.   *Teiidæ.*

PART III

A COLLECTOR'S EXPERIENCE
WITH REPTILES

By RAYMOND L. DITMARS

CURATOR OF REPTILES AT THE
NEW YORK ZOOLOGICAL PARK

FIG. 77.—The collector with some of his pets.

168

CHAPTER XVII

SNAKES AS HOUSEHOLD PETS—PREVAILING PREJUDICE
AGAINST REPTILES—THE USES OF REPTILES—WHY
SNAKES ARE FRIENDS OF THE FARMER—THE PLACE
OF REPTILES IN NATURE

THERE is probably no class of creatures less known,
more hated, and unjustly persecuted than reptiles;
but observation brings about a transformation of
ideas in the minds of their most persistent enemies.
One of the prevailing ideas is that reptiles are slimy,
and consequently loathsome. Yet among the several
thousand species of reptiles not one is slimy. In
habits they are far more cleanly than many household pets, and in the colors of their scaly bodies none
but persons of narrow-minded prejudice can fail to
admire Nature's lavish tints. The comparatively
small proportion of reptiles possessing poison-bearing fangs are not provided with these instruments
for the purpose of slaying mankind; the fangs of the
poisonous serpent are intended by Nature to be an
aid to the creature in procuring its prey.

Often is the question asked, "Of what possible
use are reptiles?" A knowledge of the food of rep-

tiles explains their use in the economy of Nature. Many species of snakes, among them the larger constrictors of the Southern States, are found more abundantly in fields of growing corn or sugar-cane than in any other location. During the spring, when these fields are furrowed, these reptiles are cast from their hiding-places by the plow, showing their aversion against leaving the fields even for the period of hibernation. Their presence there is easily explained. Coming from the near-by woods are rodents and other small creatures that collect in these fields to feed upon the products of tilled soil. Unmolested, their ravages would be disastrous; but Nature has carefully laid her plans to check their multiplication. A single blacksnake, during the summer months, will devour dozens upon dozens of mice, will prowl through the burrows of shrews and moles, devour the young, and go in search of more. The appearance of a blacksnake in a field of grain guarantees the destruction of many of the farmer's most elusive enemies.

Although a great many reptiles play havoc among the smaller mammals, there are other species which confine their attention to the regulation in numbers of their own class. These are cannibals, and feed almost exclusively upon snakes and lizards. Many of the cannibal snakes are gifted with an immunity against the poison of the venomous species, which they attack and devour.

Burrowing in the pulp of decaying trees and searching in the bark of the living, are forms of reptile life but seldom seen by the eyes of the uniniti-

ated into the wonders of natural history. Day by
day, in the pursuit of their sustenance, these creatures, some of them exquisitely beautiful in their
coloring, are waging constant warfare against the
great army of insects which must be kept in check,
or life upon this earth, of both plant and animal kind,
would be menaced by a terrible scourge.

That every existing creature, every organism, no
matter how minute or lowly, has some duty to perform upon this earth is indisputable. Nature toler-
ates no useless creatures; a race of such must rapidly
degenerate and perish. Thus does reptilian life perform its duties; and cannibalism among its own members, as well as its natural enemies of the wilds, keep
it within the bounds of Nature's plans.

CHAPTER · XVIII

THE majority of reptiles lay eggs which are left
to hatch from the heat of the sun, or by the decom-
position of vegetable matter in which they are deposited. Among the snakes there are a few exceptions to the rule of leaving the eggs unprotected.
The species comprising the genus *Python* incubate
their eggs by coiling closely about them. During
this action on the part of the female snake a strange
thing happens. The body of the ordinarily cold-
blooded reptile assumes a much higher temperature
than the surrounding atmosphere: this rise in temperature is often as great as twenty degrees. After
the period of incubation, which varies from six to
eight weeks, has passed, the serpent's body gradually
acquires the temperature of the surrounding air.
When the little pythons have once hatched they are
fully able to care for themselves, and the mother
pays no further attention to them.

The female crocodilians frequently remain in the vicinity of the eggs until they hatch, but their presence in no way appeals to the development of the young. The eggs of the crocodilians are generally deposited in heaps of decomposing vegetable matter scraped together by the mother.

Among the snakes a number of species bring forth the young alive. The poisonous snakes, excepting the cobras and allied species (the *Elapidæ*), produce the young alive. From the moment of their birth the young snakes are provided with venom-bearing fangs, and are, moreover, fully capable of using them. The characteristic of bringing forth the young alive belongs also to a large number of the harmless snakes. Species that lay eggs are called *oviparous* snakes; those that give birth to living young are scientifically termed *ovoviviparous* species. Although belonging to the same family, the *Boidæ*, the pythons of the Old World, lay eggs, while the boas, which inhabit the Western Hemisphere, produce fully developed young.

The number of young brought forth by different species of snakes varies greatly. Our common garter-snake, an ovoviviparous species, produces on an average about thirty-five young; the common water-snake occasionally gives birth to yet larger broods; on one occasion a specimen in the writer's possession gave birth to sixty thriving youngsters.

The majority of the highly poisonous snakes bring forth comparatively small broods, their number seldom exceeding fifteen. Some of the tropical vipers, however, are exceptions to this rule. The copperhead snake of this country gives birth to about eight or nine young. Highly prolific among the snakes are the giant boas and pythons, as two illustrations from the writer's personal observation will show; in one instance a captive boa (*Boa constrictor*) gave birth to sixty-four fully developed young; while a huge python deposited seventy-nine eggs, which she gathered in her coils and guarded jealously from the kindly interest of her keeper.

Reptiles live much longer than warm-blooded animals; some species have been observed to have attained astonishing ages. Longest to subsist among them are the tortoises; following in order are the crocodilians, lizards, and snakes. A giant tortoise in the Zoological Park is estimated to be over three hundred and fifty years old. Many records are on hand where tortoises have been known to live for more than a century. It may be safely said that as long as conditions remain satisfactory for the existence of reptiles they go on living indefinitely; "old age" comes to them when handicapped by some injury, or when disease has permanently destroyed the vigor of some part of their anatomy.

Among the tortoises and turtles old specimens may be distinguished in many instances by the smoothness of their shells. As some of these reptiles possess shelly coverings which abound with ridges and serrations, and specimens live many years in captivity without showing marked signs of wear, it is impossible to estimate the number of years required to produce the almost glassy smoothness of some specimens when it is considered how sluggish and inactive these creatures are.

Many snakes, received when fully adult, have been in the writer's possession for eight and ten years, and show no signs of age. Full-grown pythons have been in captivity for fifteen years and more, and finally succumbed to diseases unknown to reptiles existing in their natural conditions. It is difficult to determine the average lifetime of snakes, as previous observations of these reptiles have been so meager; but their age may be said to be considerably shorter than chelonians and crocodilians, as their growth is much more rapid.

In this connection, the writer takes some pride in describing the growth of a particularly interesting specimen. During the latter part of August a large diamond-back rattlesnake gave birth to nine little ones. On the day after their appearance in the world the little snakes were discovered busily engaged in shedding their skins. The operation is performed by all young snakes within forty-eight hours after their birth. None begins feeding until after it has been accomplished.

Of this brood of tiny rattlers, one soon became a favorite, owing to its especially pretty coloration. It was this specimen which furnished the writer with the notes which follow. At birth it measured eleven inches; like all favorites, it was provided with a name, and the name was " Rattles." Some three months after birth Rattles' eyes grew pearly and he prepared to shed his skin again.   Prior to this time his future rattle had been represented by a tiny black knob at the tip of his tail.   When Rattles was excited he vigorously shook his tail, and the little button became blurred in rapid motion, but produced no sound. Then, as he prepared to shed his skin, a new joint developed at the base of the button.   When Rattles

FIG. 78.—Rattlesnake.

crawled out of his old clothes he uncovered this new ring, which had been growing under the skin ; it soon grew dry and brittle, and the "button," fitting to it loosely, produced a faint, buzzing sound when shaken. From tiny mice, Rattles had graduated to mice of good size, for which young rats were soon substituted.

When six and a half months old Rattles shed his third skin, one having been shed shortly after birth, and the other some three months hence, as described. He now possessed "two rings and a button," which buzzed quite noisily. The snake at this time measured thirty-eight inches, and easily swallowed, entire, half-grown rats.

During all his observations of this snake, as well as the others of the brood, the writer noted, without exception, that the shedding of the skin, which took place on an average of every three months, was attended with the uncovering of a new joint of the rattle. At nine months after its birth the snake's rattle consisted of three joints and the original button.

When Rattles celebrated his first birthday he measured four and a half feet, an increase of about three and a half feet during the year. He now fed voraciously upon full-grown rats, which died within a few seconds of a stroke by his fangs; his rattle had attained the dignity of four joints and a button.

After two years from his first birthday, let us again examine Rattles. In a large, glass-fronted cage lies a magnificent rattlesnake, its head nearly as broad as a man's hand, its colors a combination of olive, yellow, and black, forming a chain of diamond markings down its back. From the center of the coil protrudes a rattle consisting of twelve rings and a tiny button; each ring, from the tip of this appendage toward the tail, is seen to be a little larger than the preceding one, illustrating the growth of the snake; the rattle is seldom used, for the snake is very tame, but when it is sounded, there is a noise like escaping steam. The length of the reptile may be estimated at six feet, while the diameter of its thickest part is fully three inches. And this is Rattles, now three years old, his meals consisting of rabbits of fair size.

From the history of this snake, it will be understood that the practise of estimating the age of a rattlesnake by counting each joint of its rattle as one year is far from correct, as the number of joints acquired annually by the captive reptile averaged four, and would probably be one, possibly two less in the case of a wild snake, owing to the time spent in hibernation, when growth practically ceases. It must be explained, however, that wild reptiles grow faster than captive specimens, no matter how thorough may be the care of the latter. By allowing three joints of the rattle for a year, the age of a snake may be gaged, *if the rattle is pointed*, and still retains the " button." When all of the joints of a rattle are of uniform size, the owner of the same has ceased growing, and the rattles of its youth have been lost through wear or injury at some indefinite time impossible to discover. In such a case it can not be ascertained how many rattles have been grown and lost, but of course demonstrates the snake to have attained maturity. When greatly angered, snakes with long rattles will shake off a number of the joints in sounding the instrument which warns the unwary of their deadly powers.

CHAPTER XIX

THE CARE OF REPTILES IN CAPTIVITY—ECCENTRICITY
OF APPETITE—FASTING OF POISONOUS SNAKES—HOW
THE BIG PYTHON WAS SAVED—CANNIBAL SNAKES
—NOVEL METHOD OF FEEDING THE KING COBRA—
MALADIES OF CAPTIVE REPTILES

COMPARED with warm-blooded creatures, the care of reptiles is eccentric in the extreme. Most important to be considered is *temperature*. All reptiles, including the great tropical serpents, flourish in a temperature of from 75° to 85° F. Species from temperate climes live well in a temperature of 70°, but none remains active or in "good feeding" in a lesser degree of heat.

Among reptiles the most difficult forms to keep in good health are the snakes. Possessed with appetites which make them unique among cold-blooded creatures, their care taxes the brain of the most experienced keeper as he strives frequently to keep valuable specimens from deliberately starving in the midst of plenty.

All snakes swallow their prey entire, aided by the elastic mechanism of their jaws. Different species kill their prey in various ways. The constricting

179

snakes coil tightly about the victim and quickly squeeze it to death; the poisonous snakes kill by a stroke from their deadly fangs; some species, provided with long, hooked teeth in the rear portion of the mouth, seize the quarry in a never-failing hold and swallow it alive, while others, for the most part the large rat snakes of tropical countries, pin their victim firmly to the ground under a portion of their body, during the process of swallowing. These latter snakes have also the habit of violently shaking their prey in much the same manner as a dog treats a rat. In a way, these different methods of feeding apply to the classification of snakes.

In captivity many snakes evince a remarkable reticence in feeding. Some species, especially the viperine poisonous snakes, often prefer to starve than to take the food offered them. Some of the tropical vipers absolutely refuse to live in captivity, and after exhibiting a remarkable vitality for six months without food, slowly and stubbornly starve themselves to death. The writer has observed instances of rattlesnakes having lived seven and eight months without other nourishment than water, yet remaining active and hostile during all this time.

With captive snakes these starving inclinations are seldom tolerated by their keepers. The food for the reptile is killed and thrust down its throat by force. Some snakes live under this treatment for years. The writer has superintended many operations where badly emaciated specimens were strengthened and finally brought to prime condition through this stuffing process. The majority of these obstinate specimens at length cast aside their stubborn desire to starve, and take food readily after a time; others positively decline to help themselves.

As the reptile house in the Zoological Park neared completion a giant snake arrived. Packed coil upon coil in a crate not more than four feet square, the monster had spent over three months without food or water. Over twenty feet long, with a pattern like a Persian rug, the big snake promised to be a most interesting specimen; as suiting her Oriental habitat we named her "Fatima," and at once started to prepare her quarters. The big cages not being ready, she was given temporary quarters in the animal shed, and provided with a tank, under which an oil stove burned steadily. In the tepid water she coiled her regal length and bathed for days.

About ten days after her arrival keeper Snyder noticed a change in her temper. She lay coiled closely in a corner, and hissed savagely when approached. It did not take long to discover that she was coiled about a mass of eggs, and there was exultation among the keepers. Six weeks passed; the time for the little pythons to appear had gone, and a sad spirit of realization dawned upon us all. Owing to the big snake having been chilled on her journey to the park, the eggs were spoiled. Still Fatima waited patiently for the appearance of her little family; she furiously resented interference, and matters became serious.

Eight weeks had passed since Fatima's arrival at

the park. She had now been fasting five months, and showed marked signs of emaciation. Moreover, her eyes had grown pearly, like little bubbles filled with smoke; her bright colors had faded, and the time to shed her skin had passed; yet the devotion to the eggs which would never hatch.

It had to be done at last. The big cages in the reptile house were ready; with several keepers, the writer covered Fatima with blankets, grasped her by the neck, and carried her to her new quarters. The monster was weak and thin; five men did the job, and it took little effort among them. Seventy-nine eggs were counted in the huge cluster which the snake had been endeavoring to incubate; these were of about the diameter of a hen's egg, but longer. But now another serious difficulty presented itself. Having neglected to shed the skin, it had dried and hardened, and speedy death threatened the proud Fatima. Her cage was filled with steam and left so for an hour or more, and after this Turkish bath the keepers entered; then began the job of " peeling " the big reptile. It took hours to complete this task; but when it was completed Fatima appeared in gorgeous apparel, her colors blending with the iridescence of the great constrictors. It now remained for her to take her food, and she was saved.

Poultry was offered repeatedly. The choicest feathered stock was introduced; but the glitter of the yellow eyes followed the motions of the keepers, who warily kept from within reach of the long, recurved teeth. Since her capture, six months before,

Fig. 79.—Forcing food down the throat of a big reptile.

14

Fatima had continually fasted, and it was decided to take desperate measures to save her life.

Seven rabbits were killed and neatly fastened together with twine, making one long string. Fatima was taken from her cage, held by six keepers, and the rabbits forced down her throat with a smooth pole. When two feet of the ten-foot pole protruded from her mouth the pole was withdrawn, leaving the rabbits inside; then the snake was placed in the cage again. This operation was repeated for eleven months, at intervals of ten days, and from the beginning Fatima grew rapidly stronger; after some weeks of the compulsory feeding the "python squad" consisted of twelve men instead of six, and all of these had their hands full as they swayed and fought to retain the mastery over twenty feet of reptilian muscle.

At last came victory. All through the time of compulsory feeding the keepers had continued to offer Fatima her regular food, hoping that she might begin taking it voluntarily. One night at dusk, as Snyder was feeding a large snake in Fatima's cage, which had been docile from the first, he was amazed to see within a foot of his face the great head and neck of the regal python, her yellow eyes scintillating brightly in the light of his lantern. All her attention seemed concentrated upon the black-tailed python, which was calmly swallowing a chicken. Seizing another fowl, Snyder threw it in front of the reptile as she slid from the big tree in the cage, and with wild excitement and joy beheld Fatima seize and devour it. Such was Snyder's exultation that he performed a wild dance on the spot, then rushed for more chickens. That first voluntary meal in captivity consisted of eleven big fowl. Never afterward did Fatima cause the slightest trouble in her feeding, and thus the "python squad" was disbanded.

In captivity the cannibal snakes frequently cause embarrassment by exhibiting appetites which threaten to cause a famine. This was illustrated in the case of the big king cobra in the reptile house. Every week this twelve-foot serpent received a five-foot black-snake, but it showed signs of growing thin under the

FIG. 80.—A blacksnake.

fare. It was clearly seen that unless more snakes were given to the reptile it could not thrive. The outlook was decidedly embarrassing; the cobra was making serious inroads among the exhibition specimens, and the blacksnake and coachwhip cage was almost empty. It was decided one day, after a large blacksnake had been killed, to stuff it to its utmost capacity with half-grown rats and frogs, and present the distended carcass to the cobra. When the blacksnake was fully prepared it more closely resembled a generously filled Christmas stocking than a serpent, and was equal to half a dozen snakes.

The cobra gravely inspected the unique morsel, and finally engulfed it entire, although the process was quite heroic. Since then the cobra has been fed " stuffed " snakes, and presents a sleek and hearty appearance.

But it is not only the cobra that has been deluded, in a spirit prompted by economy. Rat snakes are tempted to eat strips of beef by clipping a small quantity of fur from a rat or rabbit and sprinkling it over the meat. A small amount of fur will bait a dozen strips of meat. In the same way the larger snakes are induced to partake of meat by sprinkling over it a few chicken feathers.

It is a mistake to suppose that captive snakes must kill their prey or they will not eat. Of course, in a natural condition this is necessarily the case. The majority of captive specimens feed readily upon freshly killed material, and such food is always preferred by their keepers. The poisonous snakes generally offer an exception to this rule, although the cobra and its allies are not at all particular. If a live rat be placed in a cage containing a dozen ravenous snakes all rush for it at once, and a serious tangle ensues. The keepers feed the specimens individually, killing the food and offering quietly to each specimen its share of the meal. If the food were thrown at random into a cage of snakes the smaller specimens, if persistent in their hold of choice morsels, would be swallowed with them, and find a resting-place in the interior of their more powerful cage-mates.

The maladies of snakes are few, but of these few the consequences are often quick and fatal. Most prevalent among captive snakes is " canker." This disease, generally attacking the mouth-parts, is the cause of nine-tenths of the deaths among the big snakes of the shows. It is most likely to occur where conditions are not suited to the reptile, and frequently results from chilling. The big constrictors are particularly subject to canker; with them the disease is generally incurable. The term " canker " is purely a popular one, and is invariably used by the animal dealers and show people. The first sign of the disease is inflammation of the mouth-parts; sores rapidly form, throwing off small white flakes; these lesions become gangrenous and penetrate into the jaw-bones, showing microbic characteristics, and in fact resembling in a way diphtheria. The best method of treatment is the application of disinfecting or antiseptic solutions of mild character. The writer has cured many cases of " canker " by washing out the reptile's mouth twice daily with a saturated solution of boric acid.

It is not probable that snakes are troubled with

the malady when in a wild state. From the many cases observed, the writer is led to believe that the sores in the mouth are frequently brought about by the fact of the reptile's striking frequently, as wild reptiles do, and the subsequent infection of these lesions by the many forms of microbic life which abound in the quarters of captive animals, especially if these are not vigorously disinfected at frequent intervals. Snakes being naturally delicate in captivity, suffer from such infections, and, in the case of weak specimens, the blood does not possess enough of its sterilizing qualities to fight the invading germs, which start their colonies and poison the surrounding tissue.

The care of reptiles is anything but a mechanical process. Unlike the mammals and birds, their feeding is eccentric and indefinite. The keepers must be thoroughly versed in the peculiar characteristics of their charges, and sympathetic in the extreme in the filling of their many needs.

CHAPTER XX

In captivity, poisonous serpents seldom adapt themselves so readily to their surroundings as do the harmless reptiles. Venomous snakes generally retain their wild disposition and can never be trusted. While most harmless serpents submit to handling after a few weeks in captivity, their poisonous relatives resent the least attempt at familiarity by the use of their formidable fangs. For the most part these reptiles are high-strung and nervous; many feed irregularly or refuse food altogether; with the exception of those species which resemble in form the harmless serpents, like the cobras and closely related snakes, the life of the captive poisonous reptile is generally of short duration. Few of the viperine snakes live more than two or, possibly, three years, unless they be reared from an early age. In the latter case many delicate species live indefinitely.

189

There is only one species of poisonous · lizard known. This is the good-natured Gila "monster," a hardy and attractive animal for exhibition. It inhabits the burning wastes of Arizona, and delights to bask in a temperature which would kill many of the snakes. The poison of this creature, although resembling in composition that of the snakes, is by no means as powerful as with the latter. The Gila "monster" is provided with a number of grooved teeth in the lower jaw, and can bite with a power that approaches the grip of a metal vise.

In the care of a great collection it frequently becomes necessary to handle the poisonous snakes, a process as dangerous as it is simple. The reptile is coaxed into a favorable position with a long stick, when the latter is placed firmly across its head, pinning that member to the ground. The operator quickly grasps the reptile by the neck, immediately behind the head, and victory is his. Held in this position the snake can not turn and bite, although its jaws will often fly open and shut, disclosing the poison-bearing fangs in a manner quite terrifying to any one with weak nerves. Thus the poisonous snakes are taken from their cages and helped out of old skins, or relieved of a portion of their poison for the purpose of study.

While experimenting with snake poisons the writer had occasion to extract the venom from a number of water-moccasins. The apparatus used was simple, but very effective. Over the top of an ordinary graduating glass was tightly tied a piece of thin chamois.

The snake was caught by the neck, its jaws applied to the chamois, and it immediately bit fiercely, sending the fangs through the soft covering of the glass vessel. As it closed its jaws on the apparatus the fangs discharged their venom. From a snake four feet in length fully half a teaspoonful was the amount usually obtained, and the large diamond-back rattlesnakes of our Southern States are able to eject a considerably larger amount from their deadly weapons.

The entire amount of venom in the poison-sacs is never expended in a single bite; a poisonous snake of moderate dimensions is capable of dealing successively half a dozen deadly wounds. The manufacture of venom is rapid, and even though a snake's glands be entirely emptied of their contents through a mechanical process, forty-eight hours affords ample time for refilling the glands under the chemical action exerted by the organs which manufacture the deadly fluid.

That a poisonous snake may be rendered harmless by the extraction of its fangs is a fallacious idea. The venomous serpents are constantly shedding their fangs. Strangely enough, the fangs which are shed are, in the majority of instances, swallowed by the reptiles. This is frequently illustrated in the cleaning of cages. It may be explained, in a way, from the fact that a loose fang may be left embedded in the body of an animal struck by the snake for food, and subsequently swallowed, but it is nevertheless remarkable that a snake which has not eaten for months regularly sheds its fangs, and swallows them in the process.

Growing behind the two fangs in use are other fangs, and still the embryos of others. Every six or eight weeks the fangs are shed. By a wonderful provision of Nature, the serpent never loses an old fang until the new member is strongly attached at its side, and connected with the poison-gland; then the old fang comes loose from its socket, and is left deeply embedded in the body of the next animal struck for food. In case the snake were artificially deprived of its fangs, the openings of the poison-ducts would continne to discharge their secretions, and if, in such an instance, the reptile should inflict a bite, the wounds made by the small palatine teeth would form excellent sources for the absorption of the poison.

Although much has been said concerning the aggressive disposition of the venomous snakes, the writer, after many years of experience, has failed to note a single instance of the deliberate intention on the part of a poisonous serpent to pursue or exhibit aggressiveness toward an enemy. The defensive is always the attitude assumed by the poisonous reptile, and although a specimen will occasionally show the temper of a fiend, it never advances to the attack, but always keeps its corner. By no means cowardly, the reptile simply wishes to be left alone. The viperine snakes, comprising the rattlesnake, copperhead, water-moccasin, and fer-de-lance of this hemisphere, and the typical vipers of the Old World, coil themselves when assuming a fighting position, although the coil is in no way necessary for the act of striking. It merely forms an anchor, and aids the reptile in dealing a hard blow. Such snakes can strike one-half their length, and a six-foot rattlesnake may launch its terrible jaws three feet at the object of its wrath. It happens oftener, however, that the strike of a viperine snake is about a third of its length. Stories of snakes casting themselves bodily in the direction of their anger are entirely erroneous.

Owing to the great danger attending the handling of the larger venomous snakes, interesting measures are frequently called into play when it is necessary to shift specimens or to treat them for disease or injury. It was with great anxiety that we noticed a small growth appearing on the king cobra's upper jaw. The lesion grew until it resembled a small abscess, and but one thing was left to do. That was to remove the growth. As the cobra was very powerful, and represented the most deadly known species of reptile, the operation presented many difficulties. But it was successfully performed. A coachwhip snake about six feet long was killed and thrown into the cobra's cage. Fortunately the cannibal still retained a good appetite. In the meantime the instruments, consisting of a pair of surgical scissors, a rubber syringe, together with a bowl of disinfecting solution, were made ready. When the coachwhip was half-way down the cobra's throat it was grasped by the tail and pulled toward the door of the cage, bringing with it the cobra, which held on tenaciously. By twisting the coachwhip's body the cobra was rolled over on one side and the abscess in the mouth disclosed. The operation was necessarily a quick one.

There was scarcely time to cut away the diseased
tissue, flush out the wound with the syringe, and
quickly close the cage door when the cobra prepared
to disgorge the snake and fight. For a moment after
it was over, the big serpent looked surprised, then,
after due consideration, the coachwhip was swallowed.
The operator is not ashamed to acknowledge that
when the iron door rolled to and shut off the danger his pulse had quickened to a substantial degree.

Avoiding technicalities, the poison' of all snakes
may be said to be composed of the same elements;
but in different species these elements vary in their
proportions, and there are different symptoms after a
bite. The viperine snakes are provided with a venom
which is composed of about ninety per cent of a
product which acts upon the blood and ten per cent
of a nerve poison. The bite of these snakes produces great local effect. There is the destruction of
the blood itself and of its vessels, with great attend-
ant swelling of the bitten parts. In the cobras and
coral snakes these proportions are exactly reversed,
for the venom is a deadly *nerve* poison, and causes
but little local effect. This poison tends to paralyze
the nerves, the walls of the chest collapse, and there
is inability to breathe, followed by speedy death unless immediate and proper remedies be applied.

As all snake-poisons tend to paralyze the heart's
action, a stimulant is necessary and beneficial; hence
the use of strychnine as a valuable alterative, owing
to its action as an excitant. Whisky is also a good
stimulant, but the practise of taking it in great quan-

tities after snake-bite is in no way to be recommended.

The treatment of snake-bite has fallen into line with the scientific administration of an antitoxin, as in the case of many diseases, and probably most of them at no far future time. The antitoxin or serum employed is obtained by immunizing horses against the action of the venom, and procuring from these animals a product which has shown the same beneficial results as those exhibited by the serum manufactured at the laboratories of the New York City Board of Health and used so extensively in the treatment of diphtheria. This antivenomous serum is now being employed in India, where the death-rate from the bites of poisonous snakes has averaged 20,000 a year.

Constant association with the venomous snakes renders their keepers entirely immune to fear, but carelessness in a reptile house is considered a positive crime, and caution is the watchword. A trip through the reptile house at night shows much activity among the viperine snakes, while the slender cobras, which delight to bask in the genial sunlight, lie quietly sleeping. Glittering in the light of a lantern, the rattlesnakes, copperheads, and tropical vipers may be seen alert and gliding noiselessly about their cages. It is at this time that these creatures prefer to take their food.

CHAPTER XXI

COLLECTING REPTILES — HOW WATER-SNAKES ARE
CAUGHT — DIFFICULTIES IN CAPTURING LIZARDS—
HUNTING THE LOCAL REPTILES — WHERE REPTILES
MAY BE FOUND—THE TIME TO COLLECT—HUNTING
AT NIGHT

As the writer begins this chapter a flood of remi-
niscences comes to his mind, and he recalls the balmy,
humid air of the South Carolina coast in spring; the
graceful live-oaks, clad in the long, trailing moss;
a moonlit background with silhouettes of tall pal-
mettos, and many happy days which he has spent
among the savannas, for there reptilian life abounds.

To one thoroughly interested, the collecting of
reptiles is comparatively easy. The necessary appli-
ances are few, and they may be carried in a gripsack
or in the pockets of a canvas coat. It might be incidentally explained that these remarks exclude the big
reptiles of the tropics, which will be considered later.

On a windy March day keeper Snyder, of the
reptile department of the Zoological Park, and the
writer left New York in a whirl of sleet, and arrived
three days later in a land where roses bloomed in pro-
fusion. Our collecting headquarters were on the Sa-

vannah River. The collecting outfit consisted of canvas bags, plenty of fine soft copper wire for noosing, and—an abundance of quinine. On the return of the expedition to New York we brought four hundred and nine reptiles, comprising snakes and lizards.

Abounding in the Southern swamps and along the waterways running parallel with unfrequented roads, are water-snakes of many kinds, some of them exceedingly beautiful in brilliant shades of red and yellow, some sinister and ugly in coloration and equally so in temper, others (the cottonmouths) possessing highly dangerous fangs, but all primarily anxious to seek their native element when disturbed. Of all the shy creatures the writer has ever seen, these Southern water-snakes deserve first mention. It is quite disheartening to observe one moment a brilliant specimen coiled gracefully around a branch overhanging the water, and to think of the interest which the creature would cause in captivity, and an instant later to see the prize glide sinuously into the water below, leaving but a few bubbles as mementoes. This often happens after one has stalked the animal in most cautious fashion with a long bamboo pole equipped with a ready noose, while the persistent mosquitoes attack the collector viciously, seemingly realizing that if he moves a finger-tip the coveted specimen will vanish.

It is safe to say that at least half the water-snakes stalked in this manner, escape. The wire of the noose must be of about the thickness of a pin, and it *will* tremble as one reaches forward. A noose of thick wire does not respond to a pull from the operator in time to prevent the active snake from gliding through it as quickly as falling water. The art is to pass the noose over the reptile's head, then, with a sharp tug on the pole, make it a prisoner.

The gyrations of a captured water-snake are bewildering. It coils into all sorts of fantastic knots, and snaps at the pole, itself, and its shadow, and will ultimately break the wire unless overpowered. A few seconds' maneuvering enables the operator to get the reptile by the neck, unfasten the noose, and drop it into a canvas bag. Never will the writer forget one experience while noosing snakes in this manner. He was attracted by the head of what appeared to be a large turtle among some aquatic plants. As turtles can also be taken by a noose properly manipulated, he slipped a miniature lasso of wire over the creature's head and pulled; but instead of hauling in a turtle, the body of an enormous water-moccasin came thrashing ashore, its dark eyes sparkling wickedly; and its widely distended jaws disclosing a pair of fangs that sought for vengeance. The apparition was so startling that for a moment the collector was quite overcome with astonishment. This was his first "cottonmouth," and he presumes that the startled feeling was practically identical with "buck fever." But there were other moccasins in that swamp, their turtlelike heads just visible above the surface of the weedy water. The catch that morning was eleven, and .those same eleven snakes are thriving in captivity.

With the capture of most of the terrestrial snakes

the noose is out of the question. It is a question of the one making the fastest time—the snake in rushing to a place of security, or the collector in reaching the snake. If the collector wins, a quick grab is in order, which is usually followed by a burst of indignant protest on the reptile's part. To get the reptile by the neck before receiving a bite requires some skill; but in the case of a harmless snake little caution is necessary, and the prize is soon in a bag. Incidentally it may be said that poisonous snakes do not run, and their capture involves pinning down the head and grasping them by the neck. This sounds easy, and in fact it *is* easy for one who understands it, but the writer would entreat the novice to think twice before coming within striking distance of a poisonous snake.

Most elusive of capture are the lizards. In summer it is practically impossible to obtain many of the species. The movements of many species are so rapid that the eye can scarcely follow them. Fallen trees, exposed to the morning sun, are the favorite congregating places of many lizards, and here they perform their antics until disturbed, when they disappear like magic into crevices of the bark or into the surrounding vegetation, where their bright eyes watch carefully for succeeding events. By stealthily approaching their haunts, and upon locating a specimen, stalking it with outstretched hand, it may be sometimes seized, providing the movements of the collector are lightninglike in character. Generally, however, the collector is left ruefully examining a wriggling tail,

15

the owner of which is scampering off with the speed of the wind, never to stop until secure.

Spring is the proper time to collect lizards, as during this season the vegetation is sparse, and the reptiles, intent in their enjoyment of the sunshine, are less cautious. In South Carolina, where several species of showy lizards are abundant, the writer collected many specimens during the early spring months by stripping the bark from dead trees where the reptiles, which were quite inactive, had passed the winter. In the same district, a few weeks later, many lizards representing the species collected were seen, but they eluded capture in nearly every instance.

Although ponderous and exceedingly powerful in proportion to their size, alligators and crocodiles are easily caught by baiting a powerful hook with flesh, fastening it to a rope, and placing it in the lairs of the saurians. Since they are exceedingly tenacious of life and survive injuries which would immediately prove fatal to warm-blooded animals, the superficial wound made by the hook heals within a few days and causes the reptile no inconvenience. Better than this, however, is the practise of stealthily approaching these reptiles as they lie sleeping, and noosing them with a strong rope. In this fashion the thirteen-foot alligator "Big Mose," now in the reptile house of the New York Zoological Park, was taken.

Many of our local reptiles, owing to their shy and retiring habits, are difficult to discover. The

diminutive DeKay's snake and the closely allied
Storer's snake are generally found beneath flat
stones or strips of bark at the edges of woods.
These little reptiles are seldom seen abroad except
in spring or fall, when they delight to bask in the

Fig. 81.—"Big Mose," the alligator.

sunlight. Precisely the same are the habits of the
tiny and dainty ring-necked snake. These reptiles
are most frequently found by stripping the bark from
decaying trees.

In collecting reptiles it is useful for the beginner
to know that snakes or lizards are seldom found in
thick woods; in such places the collector will seldom
find anything but a few batrachians, such as salamanders and newts. Snakes prefer the borders of
woods or small clearings. Several of our local species frequent rocky places, and the borders of swamps
are the favorite lurking-places of others.

To the beginner a reptile hunt is generally most
discouraging. The anticipation is that careful search

will reveal many specimens. But such is not generally the case. In the northeastern United States a
collector may pass an entire day searching for reptiles and see not a single specimen. It is usually by
accident that one meets the larger snakes. Often
the collector returns wearily from a tramp through
seemingly the most favorable country without the
sign of a specimen, unless it be a crushed and battered
one on the roadside.

In spring the reptiles issue from their places of
hibernation, and, possessed with a spirit of sociability,
sun themselves in little colonies. As the season grows
older they scatter. Some glide into the long grass
of the meadows, others go searching through rocky
ground, and a locality previously teeming with reptile life becomes depopulated, so that the collector
can expect little luck during the summer months.
Those first warm days, always heralded by a chorus
of frogs and toads, afford the reptile hunter's most
favorable time. Along old stone walls and hedge-
rows the reptiles venture forth, and the sparse vegetation of the season makes their discovery and capture easy.

The poisonous snakes may be hunted during the
summer months along quiet roads, at dusk. When a
long spell of heat and drought has been broken by a
refreshing shower they also venture forth to hunt
their prey.

It is surprising to note the seeming rarity of the
local poisonous snakes compared with the harmless
species. Many times has the writer tramped through

miles of country said to contain rattlesnakes and copperheads, and he has seen only an occasional dead specimen by the roadside. In the extreme Southern States, however, poisonous snakes are more numerous, and rattlesnakes of several species may be hunted at night. Well does the writer remember the consternation among the colored folk created by his companion and himself during their nocturnal hunts. Nor can these simple people be blamed for evincing astonishment at the apparition of two canvas-clad figures entering the swamps at night, armed with a powerful acetylene lamp, and emerging later with canvas bags which writhed and pulsated with struggling serpents.

So difficult is it sometimes to discover specimens in good "snakey" ground, that a friend of the writer tried the novel plan of taking with him a pair of opera-glasses and surveying the bushes and grass from some elevated point. With the glasses he once discovered a young copperhead snake swallowing a wood-mouse, presumably some fifty feet from him, among some bushes, but after taking the glasses from his eyes and searching carefully for the snake he failed to find it. This happened in the fall, when the ground was well covered with dead leaves, and the gentleman declared that if the snake had crawled away he would have heard it rustling through the leaves. Several times the writer has noted the similarity of the copperhead snake to autumnal foliage, and one of the finest specimens added to his collection was discovered coiled within a short distance of his feet as he sat resting after a long walk through the woods. The writer had been facing in the snake's direction for fully a quarter of an hour, but so closely did the reptile resemble a little heap of fallen leaves as it lay coiled that it failed to attract his attention, and might have entirely escaped his notice had it not vigorously vibrated its tail as a significant warning of its presence.

CHAPTER XXII

INTELLIGENCE OF REPTILES—TRAINING ALLIGATORS—
THE STORY OF SELIMA—DO SNAKES SWALLOW
THEIR YOUNG?—GIANT TORTOISES—THE LAST SUR-
VIVORS OF THE REPTILIAN AGE

In this final chapter the writer seeks to describe
odd phases of reptile life, the intelligence of these
creatures and how it is shown, and the peculiar characteristics observed in the case of several particularly
interesting specimens.

As a rule, reptiles show no great amount of intelligence. Devoid of affection, their interest in the
person who cares for them is prompted either by appetite or hostility. With a view of experimenting
upon the intelligence of large saurians, and incidentally devising more convenient feeding measures,
a collection of big alligators at the reptile house
was put through a course of training. Instead
of having their fish and meat thrown into the big
tank, where they could devour it at leisure, the
food was offered from the edge of the tank by the
keepers. The intention was to teach them to take
their food from the men. In this way the supply of
food could be regulated, and feeding-time would

be made more interesting to visitors. For days the 'gators deliberately starved and their food went to other animals. Time and time again the keepers presented tempting morsels from the edge of the tank; five pairs of yellow eyes gleamed hungrily, but obstinacy still ruled. At last temptation proved too strong. " Big Mose " swam toward his keepers, and his cavernous mouth yawned for only a second, but long enough. A fowl was quickly cast between the gaping jaws, and the spell was broken. From that time on " Big Mose " stood ready with open mouth at feeding-time. His companions soon followed his example, and some three weeks after the beginning of the experiment the 'gators had all acquired the habit of lining up for meals, with mouths wide open, a practise which continues now.

The domestication of the 'gators was convincing enough as to the possibilities of training reptiles, but was exceeded in interest by an episode involving a snake. The snake was a python, and the reader can draw his inferences from its behavior. At a circus and menagerie visited by the writer there was found in an annex a large case guarded by a young woman. The case was enameled in white and elaborately decorated with brass, and across the front, in shining letters, was the word SNAKES. Heralded by a blare of brass and crash of drums, the Lady of the Serpents drew forth yards upon yards of glittering, richly tinted pythons. One of these she coiled about her neck and shoulders. The reptile was exceedingly beautiful in coloring and seemed especially docile. After the act of the "snake enchantress" the writer made his way to her throne and inquired about the beautiful python which had so attracted him. He was informed that the snake was called "Selima," and had been in the show business for some years. Selima was very fond of her mistress, was the information imparted, and would not eat unless fed by hand. The conversation concluded by the purchase of Selima, and the reptile was taken to the reptile house of the New York Zoological Park.

The python seemed actually to miss its old life and grow lonesome. When the keepers in their trips down the line of cages came to her compartment and rolled back the iron door, she would crawl over their shoulders, and seemed to appreciate their attention. If placed back in the cage, she would immediately crawl out again. She would never eat unless the food were given her by hand. Having become much interested in the snake, the writer took her feeding entirely in hand. Day after day she was taken from her cage for the benefit of privileged visitors as an example of reptilian docility.

Some months after Selima's installation, the writer was seriously injured. Month after month went by, but finally came recovery and a release from the sick-room. After a three-months' absence he returned to the reptile house, there to discover that Selima's cage contained a huge rattlesnake, which glared with stony and unfriendly eye at its observer. "Where is Selima?" was the immediate question. The keepers

explained that she had deliberately starved herself to
death. Not fully appreciating the peculiar habits of
the snake, they had regularly placed the food in her
compartment, and gone "down the line" to look
after their other charges. Finding she did not eat,
the snake was subjected to the vigorous process of
having food run down her throat by force; but under
this treatment she did not thrive, and finally developed the dreaded "canker."

In describing this incident, the writer does not
wish to argue that the snake deliberately starved itself
from grief, but simply desires to explain what actually
occurred. Aside from the daily inspection of the
cage, and the introduction of the customary food at
five-day intervals, Selima had received no special
attention during the writer's absence. This followed
the time when his friends were so frequently introduced to the reptile, which was taken from its cage
on each occasion; at such times, as has been explained, the snake seemed to appreciate being handled. Although he realizes that the assertion is a bold
one, the writer contends that this snake, which had
been accustomed to being noticed and handled, missed
the many attentions previously received, and also
missed the practise of feeding it by hand, and, under
the changed conditions, worried and lost appetite;
and its long fasting led to its death.

Following the story of Selima, it might be interesting to bring up that familiar query: "Do snakes
swallow their young to protect them, in time of danger?" Those who believe in this performance,

which, if possible, would tend to demonstrate great affection on the part of the parent for her offspring, are generally persons who have paid little attention to the scientific side of natural history. In all his experience, both afield and with captive specimens, the writer has failed to notice even an intimation of such an occurrence, which is practically a physical impossibility. "Observers" allege that the young run quickly down the mother's throat; just before this happens she makes a whistling sound to call them together. As snakes are deaf, this latter statement is quite absurd; furthermore, it would necessarily take some little time for a colony of young snakes to make their way down the smooth throat of the parent; and again, it can be stated that if the young snakes ever reached the interior of the parent, where gastric juices strong enough to dissolve bones and teeth are stored, they would soon be killed by these chemicals. It is reasonable to say that the observations of the alleged protecting of young snakes by the mother in this manner are the results of observers mistaking a cannibalistic reptile devouring its prey, for a fond parent "swallowing" her offspring.

Before closing these remarks about his reptilian friends, the writer desires to briefly describe a collection of creatures which can not fail to enlist the interest of all. From a far-off group of desolate islands, abounding in innumerable craters, their rocks and sand bleaching under a tropical sun, came five representa-

tives of an age that has long passed. The islands
whence these creatures came are supposed to be one
of the rare portions of this earth left undisturbed
when, thousands upon thousands of years ago, terrific
volcanic disturbances shook the globe, and the seas
rushed over whole continents while others were born
above the waters.

These five representatives of the age of reptiles
are giant tortoises from the Galapagos Islands. Of
the times when scaled and plated forms of gigantic
proportions—forms like the visions seen in troubled
dreams—stalked in abundance through an atmosphere
of humidity and heat in forests of equally gigantic
foliage, these great tortoises are the sole survivors.
Weighing three hundred and fifty pounds, the largest
specimen to arrive at the New York reptile house
makes an ordinary land tortoise appear in about the
same proportion as a musket-ball to the huge round
shot of an old-time cannon.

The largest specimen of the five which arrived
at the Zoological Park was appropriately named
" Buster." After due comparison with the few other
specimens in captivity, and records of the same, it
was decided that Buster was about three hundred and
seventy years old. During all this time he had slowly shuffled about the sterile soil of a volcanic island,
devouring cactus leaves, and growing slowly—very
slowly, probably an inch or so every five years ; then
he stopped growing, and his great shell began to wear
against the rocks. It is estimated that this wearing
must have taken a couple of centuries, as these crea-

tures are not noted for activity. When captured,
Buster took matters easy, simply blinking hard and
puffing indignantly. It took twelve men some days

FIG. 82.—A giant tortoise.
By permission of the New York Zoological Society.

to get him from volcanic ground down to the coast,
about fourteen miles away.

Without much difficulty, Buster can carry two
men on his back. His limbs resemble the extremities
of a small elephant, and are fitting illustrations of his
strength. During the summer, he and his four companions—which cost the neat sum of one thousand dol-
lars—are fed upon watermelons. Buster's share of

these dainties is generally two of the largest, including all portions from the rinds to the seeds.

Quite different are these tortoises from species ordinarily seen. On crowded days, when the fence about their enclosure is lined with visitors, they take an active interest in the spectators and stalk about, close to the fence, holding their heads erect to the utmost limit. Frequently they have short combats, snapping fiercely at each other ; and these elephantine combats suggest scenes of the Age of Reptiles. At such times they utter a shrill trumpeting sound, which can be heard for some distance. But these little quarrels are always of short duration, and never result in injury. As sundown approaches they all trudge slowly, one after another, to their favorite corner, where their keeper provides bedding of hay ; in this they turn slowly round and round, until partially concealed. By sunset, all have sought the same corner for the night—and thus we leave them, sleeping.

INDEX

213

16

(1)

THE END

APPLETONS' HOME-READING BOOKS.

D. APPLETON AND COMPANY, NEW YORK.

OTHERS IN PREPARATION.

D. APPLETON AND COMPANY, NEW YORK.

16

(1)

THE END

APPLETONS' HOME-READING BOOKS.

Edited by W. T. HARRIS, A. M., LL. D., U. S. Commissioner of Education.

The purpose of the HOME-READING BOOKS is to provide wholesome, instructive, and entertaining reading for young people during the early educative period, and more especially through such means to bring the home and the school into closer relations and into more thorough cooperation. They furnish a great variety of recreative reading for the home, stimulating a desire in the young pupil for further knowledge and research, and cultivating a taste for good literature that will be of permanent benefit to him.

Others in preparation.

D. APPLETON AND COMPANY, NEW YORK.

BOOKS FOR NATURE-LOVERS.

By F. SCHUYLER MATHEWS.

Familiar Flowers of Field and Garden.

New edition. With 12 orthochromatic photographs of characteristic flowers by L. W. Brownell, and over 200 drawings by the Author. 12mo. Cloth, $1.40 net ; postage, 18 cents additional.

Familiar Trees and their Leaves.

New edition. With pictures of representative trees in colors, and over 200 drawings from nature by the Author. With the botanical name and habitat of each tree and a record of the precise character and color of its leafage. 8vo. Cloth, $1.75 net ; postage, 18 cents additional.

Familiar Features of the Roadside.

With Illustrations by the Author. 12mo. Cloth, $1.75.

Familiar Life in Field and Forest.

With many Illustrations. 12mo. Cloth, $1.75.

Insect Life.

By JOHN HENRY COMSTOCK, Professor of Entomology in Cornell University. New edition. With full-page plates reproducing butterflies and various insects in their natural colors, and with many wood engravings by Anna Botsford Comstock, Member of the Society of American Wood Engravers. 8vo. Cloth, $1.75 net ; postage, 20 cents additional.

Plants.

Plant Relations and Plant Structures in one volume. By JOHN M. COULTER, A. M., Ph. D., Head of Department of Botany, University of Chicago. 12mo. Cloth, $1.80 net. One of the Twentieth Century Text-Books.

The Art of Taxidermy.

By JOHN ROWLEY. Finely illustrated. 12mo. Cloth, $2.00.

D. APPLETON AND COMPANY, NEW YORK.

FOR NATURE-LOVERS AND ANGLERS.

Familiar Fish: Their Habits and Capture.

A Practical Book on Fresh-Water Game Fish. By EUGENE McCARTHY. With an Introduction by Dr. David Starr Jordan, President of Leland Stanford Junior University, and numerous Illustrations. 12mo. Cloth, $1.50.

This informing and practical book describes in a most interesting fashion the habits and environment of our familiar fresh-water game fish, including anadromous fish like the salmon and sea trout. The life of a fish is traced in a manner very interesting to nature-lovers, while the simple and useful explanations of the methods of angling for different fish will be appreciated by fishermen old and young. As one of the most experienced of American fishermen, Mr. McCarthy is able to speak with authority regarding salmon, trout, ouananiche, bass, pike, and pickerel, and other fish which are the object of the angler's pursuit. The book is profusely illustrated with pictures and serviceable diagrams.

"The book compresses into a moderate space a larger amount of interesting knowledge about fish and fishing than any other volume that has appeared this season."—*Chicago Tribune.*

"It gives, in simple language and illustrations, much that it will be profitable for our boys to know before they begin to lay out their money, and much information that will be useful to them when they begin to go farther afield than their own immediate local waters."—*Outing.*

"One of the handsomest, most practical, most informing books that we know. The author treats his subject with scientific thoroughness, but with a light touch that makes the book easy reading. . . . The book should be the companion of all who go a-fishing."—*New York Mail and Express.*

D. APPLETON AND COMPANY, NEW YORK.

FRANK M. CHAPMAN'S BOOKS.

Bird Studies with a Camera.

With Introductory Chapters on the Outfit and Methods of the Bird Photographer. By FRANK M. CHAPMAN, Associate Curator of Mammalogy and Ornithology in the American Museum of Natural History. Illustrated with over 100 Photographs from Nature by the Author. 12mo. Cloth, $1.75.

Bird-Life. A Guide to the Study of our Common Birds.

Edition de Luxe, with 75 full-page lithographic plates, representing 100 birds in their natural colors, after drawings by Ernest Thompson-Seton. 8vo. Cloth, $5.00.

Popular Edition in Colors. 12mo. Cloth, $2.00 net ; postage, 18 cents additional.

Teachers' Edition. With 75 full-page uncolored plates and 25 drawings in the text, by Ernest Thompson-Seton. Also containing an Appendix with new matter designed for the use of teachers, and including lists of birds for each month of the year. 12mo. Cloth, $2.00.

Teachers' Manual. To accompany Portfolios of Colored Plates of " Bird-Life." Contains the same text as the Teachers' Edition of " Bird-Life," but it is without the 75 uncolored plates. Sold only with the Portfolios, as follows :

Portfolio No. I.—Permanent Residents and Winter Visitants. 32 plates.

Portfolio No. II.—March and April Migrants. 34 plates.

Portfolio No. III.—May Migrants, Types of Birds' Eggs, Types of Birds' Nests from Photographs from Nature. 34 plates.

Price of Portfolios, $1.25 each ; with Manual, $2.00.
The three Portfolios with Manual, $4.00.

Handbook of Birds of Eastern North America.

With nearly 200 Illustrations, 12mo. Library Edition, Cloth, $3.00; Pocket Edition, flexible morocco, $3.50.

D. APPLETON AND COMPANY, NEW YORK.

TWENTIETH CENTURY ZOÖLOGY.

Animal Life.

A First Book of Zoölogy. By President DAVID STARR JORDAN and VERNON L. KELLOGG, M. S., Professor of Entomology in Leland Stanford Junior University. 12mo. Cloth, $1.20.

"I believe it is an excellent thing, filling a gap that has long been apparent in our nature work in this country."—*Prof. Lawrence Bruner, University of Nebraska.*

"Your book is certainly an admirable discussion of biological problems up to date. It is interesting, and stimulative of thought and observation."—*Elliott R. Downing, University of Chicago.*

"The ecological treatment of zoölogy here finds a truly successful exhibition, and it is certainly very satisfactory and ahead of all previous attempts at a similar exposition for beginners in zoölogy."—*Prof. Julius Nelson, Rutgers College.*

"It is by far the best text-book on zoölogy yet published for the use of high-school students. It breathes the freshness of nature. Fortunate is the school that is permitted to use it."—*Principal W. N. Bush, Polytechnic High School, San Francisco, Cal.*

Animal Forms.

By President DAVID STARR JORDAN and HAROLD HEATH, Ph.D., Professor of Zoölogy in Leland Stanford Junior University. 12mo. Cloth, $1.10.

"Animal Forms" deals similarly with animal morphology, structure and life processes, from the lowest, simplest, one-celled creations to the highest and most complex. The two complete a full year's work in zoölogy.

The first chapter defines zoölogy, and explains minutely the morphology of a typical animal. The second chapter discusses cells and protoplasm, and prepares the pupil for an intelligent and logical study of the general subject.

In simplicity of style, in correctness of scientific statement, in profuseness and perfectness of illustration, these books are without a peer. A Laboratory Manual is in preparation. Teachers' Manuals free.

D. APPLETON AND COMPANY, NEW YORK.

APPLETONS' WORLD SERIES.

A New Geographical Library.

Edited by H. J. MACKINDER, M. A., Student of Christ Church, Reader in Geography in the University of Oxford, Principal of Reading College. Each, 8vo. Cloth.

The series will consist of twelve volumes, each being an essay descriptive of a great natural region, its marked physical features, and the life of the people. Together, the volumes will give a complete account of the world, more especially as the field of human activity.

NOW READY.

Britain and the British Seas. By the EDITOR. With numerous Maps and Diagrams. $2.00 net ; postage, 19 cents additional.

The Nearer East. By D. G. HOGARTH, M. A., Fellow of Magdalen College, Oxford ; Director of the British School at Athens ; Author of "A Wandering Scholar in the Levant." $2.00 net ; postage, 17 cents additional.

IN PREPARATION.

CENTRAL EUROPE. By Dr. JOSEPH PARTSCH, Professor of Geography in the University of Breslau.

INDIA. By Sir T. HUNGERFORD HOLDICH, K. C. I. E., C. B., R. E., Superintendent of Indian Frontier Surveys ; author of numerous papers on Military Surveying and Geographical subjects.

SCANDINAVIA AND THE ARCTIC OCEAN. By Sir CLEMENTS R. MARKHAM, K. C. B., F. R. S., President of the Royal Geographical Society.

THE RUSSIAN EMPIRE. By Prince KROPOTKIN, author of the articles "Russia" and "Siberia" in the *Encyclopædia Britannica*.

AFRICA. By J. SCOTT KELTIE, Secretary of the Royal Geographical Society ; Editor of *The Statesman's Year-Book;* Author of "The Partition of Africa."

THE FARTHER EAST. By ARCHIBALD LITTLE.

WESTERN EUROPE AND THE MEDITERRANEAN. By ELISÉE RECLUS, author of the "Nouvelle Géographie Universelle."

AUSTRALASIA AND ANTARCTICA. By Dr. H. O. FORBES, Curator of the Liverpool Museum, late Curator of the Christ Church Museum, N. Z. ; Author of "A Naturalist's Wanderings in the Eastern Archipelago."

NORTH AMERICA. By Prof. ISRAEL COOK RUSSELL, M.S., C. E., LL. D., Professor of Geology in the University of Michigan ; author of numerous works on geological and physiographical subjects.

SOUTH AMERICA. By JOHN CASPER BRANNER, Ph. D., LL. D., Professor of Geology, and sometime Vice-President Leland Stanford Junior University ; author of many publications on Brazil, Geology, and Physical Geography.

Maps by J. G. BARTHOLOMEW.

D. APPLETON AND COMPANY, NEW YORK.

A MAGNIFICENT WORK.

The Living Races of Mankind.

By H. N. HUTCHINSON, B A., F. R. *G*. S., F. *G*. S.; J. W. GREGORY, D. Sc., F. *G*. S.; and R. LYDEKKER, F. R. S., F. *G*. S., F. Z. S., etc.; Assisted by Eminent Specialists. A Popular Illustrated Account of the Customs, Habits, Pursuits, Feasts, and Ceremonies of the Races of Mankind throughout the World. 600 Illustrations from Life. One volume, royal 8vo. $5.00 net ; postage, 65 cents additional.

The publication of this magnificent and unique work is peculiarly opportune at this moment, when the trend of political expansion is breaking down barriers between races and creating a demand for more intimate knowledge of the various branches of the human family than has ever before existed.

Mr. H. N. Hutchinson is the general editor ; he is well known as a fertile writer on anthropological subjects, and has been engaged for several years in collecting the vast amount of material (much of it having been obtained with great difficulty from remote regions) herewith presented. The pictures speak for themselves, and certainly no such perfect or complete series of portaits of living races has ever before been attempted. The letter-press has been prepared so as to appeal to the widest public possible.

In "The Living Races of Mankind" attention is confined to a popular account of the existing peoples of the world. The various authors describe how the widely scattered members of the great fraternity live ; what they do ; their habitations, dress, ornaments, and weapons ; their religious feasts, ceremonies, and superstitions ; their general characteristics, manners, and customs in their daily relations with one another.

The superb illustrations are a new departure, and form a most important feature. They are accurate and beautiful reproductions of photographs from life and they form a collection not likely to be secured again in the course of the coming century.

D. APPLETON AND COMPANY, NEW YORK.

UNCLE SAM SERIES.

Popular Information for the Young Concerning our Government.

A MOST APPROPRIATE HOLIDAY OR VACATION GIFT.

Our Country's Flag and the Flags of Foreign Countries.

By EDWARD S. HOLDEN. Illustrated with 10 colored Plates. Cloth, 80 cents.

This book is a history of national flags, standards, banners, emblems, and symbols. The American flag is presented first, because every American child should know how the flag of his country came to be what it is, and that it has always been the flag of a *country*, not the personal standard of a king or of an emperor.

Our Navy in Time of War.

By FRANKLIN MATTHEWS. Cloth, 75 cents.

The leading events of our navy's achievements and their special significance are related in this book, which is designed to interest the young reader in historical research No more stirring chapter in our country's history could be selected than is contained in the story of the navy from 1861 to 1898.

Uncle Sam's Secrets.

A Story of National Affairs for the Youth of the Nation By O. P. AUSTIN. 75 cents.

This volume furnishes to the youth of the land some facts about the administrative affairs of the nation—the Post-Office, Treasury, Mint, and other functions. Especially useful to the rising generation in stimulating a desire to become better informed of the affairs of their country, and to love and reverence its institutions.

Uncle Sam's Soldiers.

By O. P. AUSTIN. 75 cents.

The purpose of this story, like the preceding, is instruction, though here it is confined to military matters, including the organization and handling of armies. The story, which purports to be the experience of two boys verging upon manhood who served in Cuba, Porto Rico, and the Philippines, gives the facts regarding modern military methods in a way that can not fail to interest.

Special Gift Edition. 4 vols., 12mo. Colored Illustrations. Bound in Handsome Red Cloth, Boxed, $3.50.

D. APPLETON AND COMPANY, NEW YORK.

BOOKS BY JOHN M. COULTER, A. M., Ph. D.,

Head of Department of Botany, University of Chicago.

Plant Relations. A First Book of Botany. 12mo. Cloth, $1.10.

"Plant Relations" is the first part of the botanical section of Biology, and, as its title indicates, treats what might be termed the human interests of plant life, the conditions under which plants grow, their means of adaptation to environments, how they protect themselves from enemies of various kinds in their struggle for existence, their habits individually and in family groups, and their relations to other forms of life—all of which constitute the economic and sociological phases of plant study.

Plant Structures. A Second Book of Botany. 12mo. Cloth, $1.20.

This volume treats of the structural and morphological features of plant life and plant growth. It is intended to follow "Plant Relations," by the same author, but may precede this book, and either may be used independently for a half-year's work in botanical study. "Plant Structures" is not intended for a laboratory guide, but a book for study in connection with laboratory work.

Plant Studies. An Elementary Botany. 12mo. Cloth, $1 25.

This book is designed for those schools in which there is not a sufficient allotment of time to permit the development of plant Ecology and Morphology as outlined in "Plant Relations" and "Plant Structures," and yet which are desirous of imparting instruction from both points of view.

Plants. A Text-Book of Botany. 12mo. Cloth, $1.80.

Many of the high schools as well as the smaller colleges and seminaries that devote one year to botanical work prefer a single volume covering the complete course of study. For their convenience, therefore, "Plant Relations" and "Plant Structures" have been bound together in one book, under the title of "Plants."

An Analytical Key to some of the Common Wild and Cultivated Species of Flowering Plants. 12mo. Limp cloth, 25 cents.

An analytical key and guide to the common flora of the Northern and Eastern States, as its title indicates. May be used with any text-book of botany.

D. APPLETON AND COMPANY, NEW YORK.

RICHARD A. PROCTOR'S WORKS.

OTHER WORLDS THAN OURS: The Plurality of Worlds, Studied under the Light of Recent Scientific Researches. With Illustrations, some colored. 12mo. Cloth, $1.75.

CONTENTS.—Introduction.—What the Earth teaches us.—What we learn from the Sun.—The Inferior Planets.—Mars, the Miniature of our Earth.—Jupiter, the Giant of the Solar System.—Saturn, the Ringed World.—Uranus and Neptune, the Arctic Planets.—The Moon and other Satellites.—Meteors and Comets : their Office in the Solar System.—Other Suns than Ours.—Of Minor Stars, and of the Distribution of Stars in Space.—The Nebulæ : are they External Galaxies ?—Supervision and Control.

OUR PLACE AMONG INFINITIES. A Series of Essays contrasting our Little Abode in Space and Time with the Infinities around us. To which are added Essays on the Jewish Sabbath and Astrology. 12mo. Cloth, $1.75.

CONTENTS.—Past and Future of the Earth.—Seeming Wastes in Nature.—New Theory of Life in other Worlds.—A Missing Comet.—The Lost Comet and its Meteor Train.—Jupiter.—Saturn and its System.—A Giant Sun.—The Star Depths.—Star Gauging.—Saturn and the Sabbath of the Jews.—Thoughts on Astrology.

THE EXPANSE OF HEAVEN. A Series of Essays on the Wonders of the Firmament. 12mo. Cloth, $2.00.

CONTENTS.—A Dream that was not all a Dream.—The Sun.—The Queen of Night.—The Evening Star.—The Ruddy Planet.—Life in the Ruddy Planet.—The Prince of Planets.—Jupiter's Family of Moons.—The Ring-Girdled Planet.—Newton and the Law of the Universe.—The Discovery of Two Giant Planets.—The Lost Comet.—Visitants from the Star Depths.—Whence come the Comets ?—The Comet Families of the Giant Planets.—The Earth's Journey through Showers.—How the Planets Grew.—Our Daily Light.—The Flight of Light.—A Cluster of Suns.—Worlds ruled by Colored Suns.—The King of Suns.—Four Orders of Suns. —The Depths of Space.—Charting the Star Depths.—The Star Depths Astir with Life.—The Drifting Stars.—The Milky Way.

THE MOON: Her Motions, Aspect, Scenery, and Physical Conditions. With Three Lunar Photographs, Map, and Many Plates, Charts, etc. 12mo. Cloth, $2.00.

CONTENTS.—The Moon's Distance, Size, and Mass.—The Moon's Motions.—The Moon's Changes of Aspect, Rotation, Libration, etc.—Study of the Moon's Surface.—Lunar Celestial Phenomena.—Condition of the Moon's Surface.—Index to the Map of the Moon.

LIGHT SCIENCE FOR LEISURE HOURS. A Series of Familiar Essays on Scientific Subjects, Natural Phenomena, etc. 12mo. Cloth, $1.75.

New York: D. APPLETON & CO., 72 Fifth Avenue.

THE LIBRARY OF USEFUL STORIES.

Illustrated. 16mo. Cloth, 35 cents net per volume;
postage, 4 cents per volume additional.

NOW READY.

The Story of Animal Life. By B. LINDSAY.

The Story of the Art of Music. By F. J. CROWEST.

The Story of the Art of Building. By P. L. WATERHOUSE.

The Story of King Alfred. By Sir WALTER BESANT.

The Story of Books. By GERTRUDE B. RAWLINGS.

The Story of the Alphabet. By EDWARD CLODD.

The Story of Eclipses. By G. F. CHAMBERS, F. R. A. S.

The Story of the Living Machine. By H. W. CONN.

The Story of the British Race. By JOHN MUNRO, C. E.

The Story of Geographical Discovery. By JOSEPH JACOBS.

The Story of the Cotton Plant. By F. WILKINSON, F. G. S.

The Story of the Mind. By Prof. J. MARK BALDWIN.

The Story of Photography. By ALFRED T. STORY.

The Story of Life in the Seas. By SYDNEY J. HICKSON.

The Story of Germ Life. By Prof. H. W. CONN.

The Story of the Earth's Atmosphere. By DOUGLAS ARCHIBALD

The Story of Extinct Civilizations of the East. By ROBERT ANDERSON, M. A., F. A. S.

The Story of Electricity. By JOHN MUNRO, C. E.

The Story of a Piece of Coal. By E. A. MARTIN, F. G. S.

The Story of the Solar System. By G. F. CHAMBERS, F. R. A. S.

The Story of the Earth. By H. G. SEELEY, F. R. S.

The Story of the Plants. By GRANT ALLEN.

The Story of "Primitive" Man. By EDWARD CLODD.

The Story of the Stars. By G. F. CHAMBERS, F. R. A. S.

OTHERS IN PREPARATION.

D. APPLETON AND COMPANY, NEW YORK.

ing the following may better represent Dr. Günther's scientific views of snakes, and help us a little into seeing their real structural peculiarities:

A. Teeth in one jaw only. Small burrowers; eyes rudimentary.
                                     (1) *Hopoterodonts.*
AA. Teeth in both jaws.
    B. None of the teeth toward the front grooved or perforated into a poison-fang. (2) *Non-poisonous Colubriforms.*
    BB. Some forward teeth so converted into poison-fangs.
        C. Fang always erect, not capable of lying back in a groove of the upper jaw. (3) *Poisonous Colubriforms.*
        CC. Fang lying back when the mouth is closed, erected when it is opened. (4) *The Viper-forms.*

The second group contains many very different families. In it are the rough-tailed burrowers, the fresh-water haunters, the egg-eaters, the tree-climbers, the whip-snakes, and the sand-snakes. These last begin by having rudiments of hind limbs to show their probable close relationship to the great boa family—the pythons, the boas, and the anaconda. All have stumps of hind legs and differ in structure mostly by the arrangement of the teeth.

The third group holds the *Elapidæ* (the beautiful poisonous tree-snakes of South America) and the sea-snakes just noted; also all the terrible serpents of India, including the cobra. All these have fangs which are *always* erect.

In the fourth group are all the vipers, rattle-snakes, copperheads, and " water moccasins " or " cottonmouths." Many are Old World or Australian, but America only has the last three; and in the Northeast United States these three are all the poi-

(1) *Burrowing snakes*, with short, round bodies, blunt tails, defective eyes, scant teeth, and no neck. They rarely appear on the surface, and are non-poisonous.

(2) *Ground-snakes*, which have bodies not out of the usual, with neck nearly always of different size from the head. They are the snakes *we* know usually. Most poisonous snakes belong here, but many are harmless.

(3) *Tree-snakes*, which usually live in trees in warm countries and have slim, whiplike tails of great length. The scales under the body often have keels or ridges, to prevent slipping sidewise. Some of these are very poisonous, as the various forms of *Elaps* in South America.

(4) *Fresh-water snakes*, which live mostly in or about the water, coming on land at times. They have the nostrils closed by valves at the upper edge of the tip of the snout. Tails round. Non-poisonous.

(5) *Salt-water* or *sea-snakes*, which live in the ocean—only one genus coming rarely on land. They can not move well on it. Their tails are flat like an eel's, and the fin is supported with the spines of the back-bone as were those of the old sea-serpents (*Pythonomorpha*) already noted. The young are born in the ocean. They are very deadly—feeding on fish which they pursue and poison before swallowing.

While this arrangement is good so far, it is not based on structure or real kinship. In a crude group-